THE CARPENTERS, CABINET MAKERS,
AND
GILDERS' COMPANION;

CONTAINING

Rules and Instructions

IN THE

ART OF CARPENTRY, JOINING, CABINET MAKING, AND GILDING;

VENEERING, INLAYING, VARNISHING AND POLISHING, DYING AND STAINING WOOD, IVORY, &c.,

THE BEST METHODS OF PREPARING

GLUE, CEMENTS, AND COMPOSITIONS,

AND A VARIETY OF VALUABLE RECEIPTS;

WITH ILLUSTRATIONS SHOWING THE VARIOUS METHODS
OF
DOVE-TAILING, MORTICE AND TENONDING,
&c. &c. &c.

BY F. REINNEL.
ARCHITECT AND SURVEYOR.

Copyright © 2018 Read Books Ltd.
This book is copyright and may not be
reproduced or copied in any way without
the express permission of the publisher in writing

British Library Cataloguing-in-Publication Data
A catalogue record for this book is available from
the British Library

Wood Finishing

Wood finishing refers to the process of refining or protecting a wooden surface, especially in the production of furniture. Along with stone, mud and animal parts, wood was one of the first materials worked by early humans. There are incredibly early examples of woodwork, evidenced in Mousterian stone tools used by Neanderthal man, which demonstrate our affinity with the wooden medium. The very development of civilisation is linked to the advancement of increasingly greater degrees of skill in working with these materials. Although it may at first seem a relatively small genre of the 'woodworking canon', wood finishing is an integral part of both structural and decorative aspects of wood work.

Wood finishing starts with sanding, either by hand (typically using a sanding block or power sander), scraping, or planing. When planing, a specialised 'hand plane' tool is required; used to flatten, reduce the thickness of, and impart a smooth surface to a rough piece of lumber or timber. When powered by electricity, the tool may be called a *planer,* and special types of planes are designed to be able to cut joints or decorative mouldings. Hand planes are generally the combination of a cutting edge, such as a sharpened metal plate, attached to a firm body, that when moved over a wood surface, take up relatively uniform shavings. This happens because of the nature of the body riding on the 'high spots' in the wood, and also by providing a relatively constant angle to the cutting edge, render the planed surface very smooth.

When finishing wood, it is imperative to first make sure that it has been adequately cleaned, removing any dust, shavings or residue. Subsequently, if there are any obvious damages or dents in the furniture, wood putty or filler should be used to fill the gaps. Imperfections or nail holes on the surface may be filled using wood putty (also called plastic wood; a substance commonly used to fill nail holes in wood prior to finishing. It is often composed of wood dust combined with a binder that dries and a diluent (thinner), and sometimes, pigment). Filler is normally used for an all over smooth-textured finish, by filling pores in the wood grain. It is used particularly on open grained woods such as oak, mahogany and walnut where building up multiple layers of standard wood finish is ineffective or impractical.

Grain fillers generally consist of three basic components; a binder, a bulking agent and a solvent. The binder is wood finish, and in the case of oil-based fillers is typically a blend of oil and varnish. The type of binder then influences the type of solvent used; oil-based fillers usually use mineral spirits, while water-based fillers (unsurprisingly) use water. Both types of filler use silica as a bulking agent as it resists shrinking and swelling in response to changes in temperature and humidity. Once the wood surface is fully prepared and stained (or bleached), the finish is applied. It usually consists of several coats of wax, shellac, drying oil, lacquer, varnish, or paint, and each coat is typically followed by sanding. Finally, the surface may be polished or buffed using steel wool, pumice, rotten stone or other materials, depending on the shine desired. Often, a final coat of wax is applied over the finish to add a degree of protection.

There are three major types of finish:

> Evaporative - i.e. wax, because it is dissolved in turpentine or petroleum distillates to form a soft paste. After these distillates evaporate, a wax residue is left over.
>
> Reactive - i.e. white spirits or naptha, as well as oil varnishes such as linseed oil - they change chemically when they cure, unlike evaporative finishes. The solvent evaporates and a chemical reaction occurs causing the resins to undergo a change. Linseed oil cures by reacting with oxygen, but does not form a film.
>
> Coalescing - i.e. Water based finishes; a combination of evaporative and reactive finishes, essentially emulsions with slow-evaporating thinners.

French polishing (an evaporative finish) is one of the most widely practiced, and highly respected wood finishing techniques, as it results in a very high gloss surface, with a deep colour and chatoyancy. The technique of applying shellac by rubbing it onto the furniture with a 'rubber' is widely regarded to have begun in France in the early 1800's, hence the description 'French Polish'. This procedure consists of applying many thin coats of shellac, dissolved in alcohol using a rubbing pad lubricated with oil. The rubbing pad is made of absorbent cotton or wool cloth wadding inside a square piece of fabric (usually soft cotton cloth) and is commonly referred to as a *fad*, also called a rubber, or *muñeca*, Spanish for 'rag doll'. It should be noted however, that 'French polish' is a process, not a material. The main material is shellac, although there are several other shellac-based finishes, not all of which class as French polishing. 'Lac' is a natural substance that is the secretion of the Lac insect 'Laccifer lacca', which is found on certain tress principally in the provinces of India and Thailand. The protective coating secreted by the lac insect is a yellow to reddish resin, which is heated, then purified and dried into sheets or flakes for commercial use.

In the Victorian era, French polishing was commonly used on mahogany and other expensive woods. It was considered the best finish for fine furniture and string instruments such as pianos and guitars. The process was very labour intensive though, and many manufacturers abandoned the technique around 1930, preferring the cheaper and quicker techniques of spray finishing nitrocellulose lacquer and abrasive buffing. In Britain, instead of abrasive buffing, a fad of 'pullover' is used in much the same way as traditional French polishing. This slightly melts the sprayed surface and has the effect of filling the grain and burnishing at the same time to leave a 'French polished' look.

Ammonia fuming (reactive) is another traditional process, but in this case, used primarily for darkening and enriching the colour of white oak. Ammonia fumes react with the natural tannins in the wood and cause it to change colour, as well as bringing out the grain pattern; the resultant product known as 'fumed oak'. The process specifically consists of exposing the wood to fumes from a strong aqueous solution of ammonium hydroxide, and can be used on other species of wood, although they will not darken to such an extent. The introduction of the method is usually associated with the American furniture maker, design leader and publisher, Gustav Stickley at the beginning of the twentieth century, but fuming was certainly known in Europe some time before this.

As is evident from this incredibly brief overview of wood finishing techniques, it is an incredibly varied and exciting genre of professional trade and individual art; a traditional craft, still relevant in the modern day. Woodworkers range from hobbyists, individuals operating from the home environment, to artisan professionals with specialist workshops, and eventually large-scale factory operations. We hope the reader is inspired by this book to create and finish some woodwork of their own.

DOVE-TAILING.

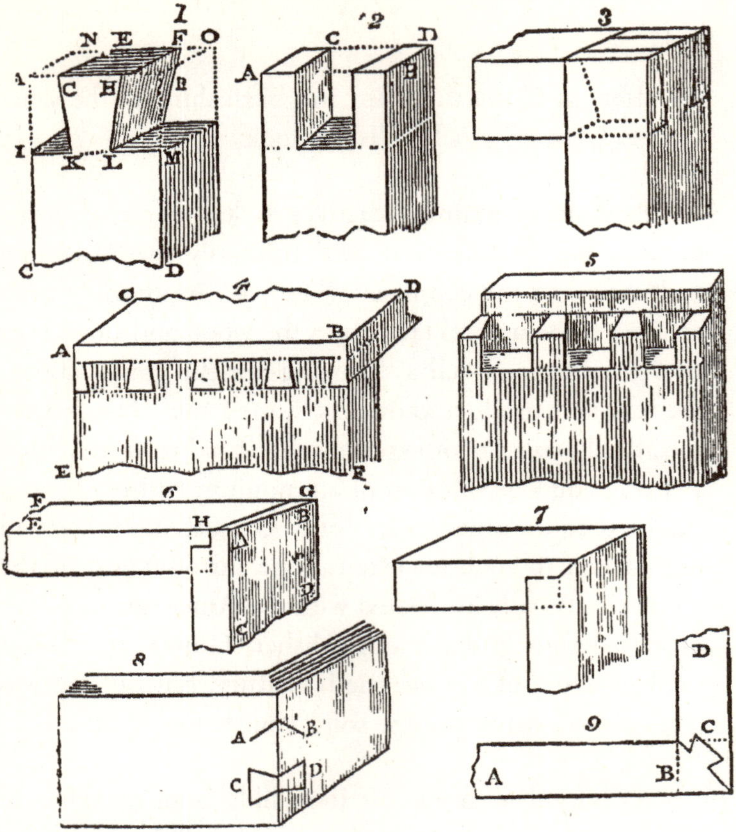

MORTICE AND TENONDING.

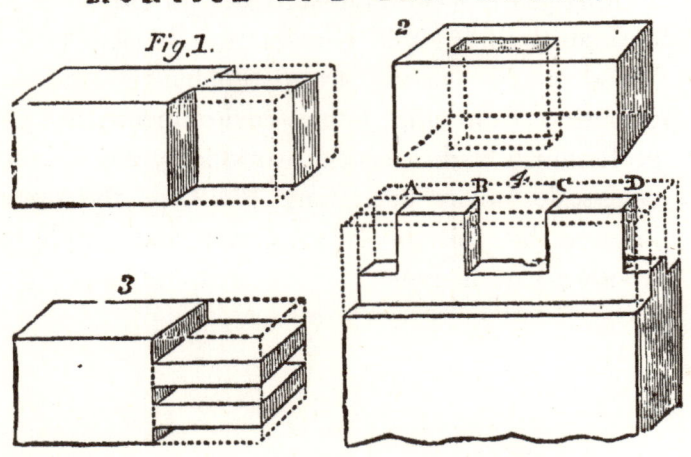

INDEX.

PART I.—CARPENTRY.

	Page
General Directions for Seasoning and Preserving Timber	9
Charring Timber	12
Another Method	13
A Composition for Weather Boarding, Paling, &c.	13
Another	14
A superior Composition for the same purpose	14
A Composition for Preserving the Joints of Framing exposed to the action of the Weather	14
Another	14
To Render Wood Incombustible	14
To Preserve Iron Work, as Bolts, Straps, &c.	15

PART II.—JOINERY.

	Page
Polish for Wainscot Work	18
Oil for Wainscot Work	19
Another Oil to Heighten the Colour	19
Polishing Oil for Mahogany	20
To Clean up Deal Panels, &c	20
To make Glass or Sand Paper	20
Stone Paper	21
Glue	21
Glue to hold against Fire or Water	23
To make a very strong Glue	24
To Glue Joints	24
To make Cement for Stopping Holes and Flaws in Wainscot, &c.	25
Another Cement much better	25
Of the different Methods of Joining Timber	25
Dove-tailing	26
Mortice and Tenonding	29
Grooving and Lapping	31
Bending and Glueing up	32
Scribing	33
Finishing of Joiner's Work	34

PART III.—CABINET MAKING.

General Remarks on Designs for Cabinet Work	37
Coloured Woods, Metals, &c.	39
Framing	40
Veneering, Banding, &c.	42
To Raise Old Veneers	43
Inlaying, &c. &c.	44
Inlaying with Shaded Wood	47
To Imitate Inlaying of Silver Strings, &c.	48
A Glue for Inlaying Brass or Silver Strings, &c.	48
To Polish Brass Ornaments Inlaid in Wood	49
To Wash Brass Figures over with Silver	49
To Gild Metal	49
Carving, Reeding, &c.	49
Moulding Ornaments, Figures, &c. in Imitation of Carving	50

DYING.

Fine Black 53	Bright Red............55
Another 54	Another 56
Fine Blue 54	Purple................56
Fine Yellow 55	Orange 56
Bright Yellow 55	Silver Grey...........57
Bright Green 55	Liquid for Brightening and
Another 55	Setting Colours 57

STAINING.

To make Imitation Rosewood 58	Black Stain for immediate use 60
Brown Vein Stain, or Imitation of Rosewood 58	To Imitate King or Botanybay Wood 61
Brown Stains to Imitate Mahogany 58	Red Stain for Bedsteads and Common Chairs 61
To Stain Beech a Mahogany Colour 59	To Improve the Colour of any Stain..............61
To give any Close-grained Wood the appearance of Mahogany 59	To Stain Horn in Imitation of Tortoiseshell 61
To take Ink out of Mahogany...................59	How to Weld Tortoiseshell 62
Easy Method of Darkening Mahogany 60	To Stain Ivory or Bone Red 62
Imitation of Ebony 60	Ditto ditto Black 62
Black Stain, or Imitation of Ebony 60	Ditto ditto Green 63
	Ditto ditto Blue 63
	Ditto ditto Yellow 63
	To Soften Ivory...........63
	To give Wood a Gold, Silver, or Copper Lustre 63

POLISHING.

The French Method of Polishing................64	Bright Polish............67
Cheap Oil Polish 64	Strong Polish............68
To Polish Ivory 65	Directions for Cleaning and Polishing Old Furniture 68
To Polish any work of Pearl 65	To take Bruises out of Furniture 69
To Polish Tortoiseshell or Horn 65	To make Furniture Paste..69
French Polishing 65	Polishing Paste 69
The true French Polish ..66	Another, for Light Coloured Woods 69
French Polish 66	Furniture Oil............70
German Polish 66	Another 70
An improved Polish 67	
Waterproof Polish........67	

VARNISHING.

Turpentine 71
Varnish for Furniture 71
An excellent Varnish for Cabinet Work..........71
To make Gold Varnish....72
A Varnish for Wood that will resist Boiling Water 72
To Varnish a piece of Furniture 73
To Polish Varnish........73
To keep Brushes in order 74

PART IV.—GILDING.

The materials to be provided with 75
Size for Oil Gilding 75
A Size for preparing Frames 76
To prepare Frames or Woodwork 76
Polishing..................76
Gold Size 77
Another 77
To Prepare your Frames for Gilding 77
Laying on the Gold 78
Burnishing................79
Matting or Dead Gold ...79
Finishing................79
To make Shell Gold 80
An excellent Receipt to Burnish Gold Size 80
To Clean Oil Paintings....80
Gold Varnish for Leather..80
To Gild Leather for Bordering Doors, Folding Screens, &c. 81
To Gild the Borders of Leather Tops of Library Tables, Work Boxes, &c. 81
To make Paste for Leather work 82

To Paint Sail Cloth, &c. ...82
Oil Gilding 83
To Gild Oil Painted Work..84
To Imitate Mahogany 84
To Imitate Wainscot 84
To Imitate Satin Wood ...84
To Stain Musical Instruments:—Crimson—Purple—Fine Black—
Fine Green — Bright Yellow....................85
Curious mode of Silvering Ivory 86
Instructions for repairing Paintings:— Damaged Surface — Repairing — Lining and Transferring .86
To Clean Tapestry, &c. ...89

PREFACE.

In issuing another edition of the "*Carpenters', Joiners', and Cabinet Makers' Companion,*" the author avails himself of the present opportunity of thanking his numerous subscribers and the trades in general, for the favourable opinions and support bestowed thereon, and trusts that the present and future efforts to render the work of permanent practical utility to those who consult its pages, to meet with a continuance of that support it has so long received.

F. REINNEL.

Walthamstow.

THE
CARPENTERS, JOINERS,
CABINET MAKERS,
AND GILDERS' COMPANION.

PART I.—CARPENTRY.

THE department of the carpenter is distinguished from that of the joiner, in that it regards the substantial parts of the edifice, such as the framing of roofs and partitions, and, in fact, all that contributes to the solidity and strength of a building; while that of the joiner is more particularly directed to the convenience and ornamental parts: it is hence evident that the carpenter should be well acquainted with the strength and stress of the materials he uses, which are for the most part in great masses; hence he should be careful not to overload his work with timber of greater magnitude than is absolutely necessary, though at the same time he should study economy in their use; he should also be able to ascertain the dimensions necessary, without weakening the building, and endeavour, by steering a mean course, to produce the maximum of strength, without overloading the several parts of his frames with too much timber; or, by putting too little, endanger the safety of the whole fabric. Thus the art of carpentry depends greatly on these two considerations, viz. the strength of the materials, and the stress or strains they are subject to; these strains may be thus—*first*, laterally, as when a tenon breaks off, or a rafter gives way close to the wall; *secondly*, the strain may be in its length, when it is drawn down or pushed in its length as a king post, where the strain tends to separate the fibre of the wood by pulling; hence, knowing the strength of the timber to resist such efforts, we are enabled to calculate the dimension necessary to be given it, to resist the probable stress it may have to overcome; all tie-beams are subject to this strain: *thirdly*, the strain may act by pressure, or the timbers may be compressed in length; this is the case with all pillars, posts, or struts; and in this case, according to the length of the post, &c., it must be of a sufficient diameter to resist the weight which it has to support. Strains are often compound, as a joist or lintel when placed horizontal, or obliquely as in a rafter; or many of those strains may be combined, as in circular or crooked work. All these things are to be particularly attended to.

The various kinds of timber used amongst carpenters are as follow:—Oak, fir, elm, ash, chesnut, and beech. A few observations on each will, perhaps, guide him in his selection to appropriate those to the several parts of a building to which they are best adapted; and, first, of OAK:—It is not necessary here to describe this timber, as every carpenter must be acquainted with it; we shall, therefore, only observe, that it is the most ponderous, as well as the hardest grain and firmest texture of any timber used in building; but from its great weight, and difficulty of working, it is but rarely used, except where very great strength is required. In selecting it, however, we should be careful that it is well-seasoned, as it is very apt to warp or cast; and the workman should be careful not to let any of the sappy parts be used, as they soon decay, and consequently are unfit for the purposes where this timber is required. The next in order is FIR, and this is the most generally useful of any timber for buildings, as we can procure scantlings of much greater dimensions than any other, and it has many advantages, as with a considerable degree of strength it combines a lightness of texture that is highly advantageous; and, also, that it is not subject to cast so much as most other timbers; it possesses a considerable degree of elasticity, and will bend a great deal before its strength becomes perceptibly impaired, consequently very proper for framing of roofs, floorings, &c. and the grain in general runs straight, and where it is sheltered from the vicissitudes of wet and dry, or protected from the weather, it will last a very considerable time; but for those parts of a building that are exposed to the weather, as window frames or cills, door posts, or any other part where the rain has access, oak is by far the best material, and will endure much longer than any other timber; and for piles, and such like, that are constantly under water, it in time gets as hard as ebony, and becomes in appearance like ebony itself; a striking instance of which, we have in those piles called Conway stakes, driven into the bed of the Thames, near Chertsey, in Surrey, supposed in the time of Julius Cæsar, which are of sufficient hardness to be used for purposes to which ebony and other hard woods are applied, such as the stocks and blades of squares, and the heads of gauges.

ELM is a wood but little used in building, being so very liable to twist and warp; it is chiefly used as weather boarding for barns, &c., and in that situation it is found to resist, (perhaps as well, or better, than any other wood,) the alternate heats of the summer sun, and the rains of a wet winter, particularly when coated with common paint or other composition.

Ash is a very tough wood, but little used in building, more, perhaps, from the demand among wheelrights and millwrights, than from any inferiority to many other woods, as it is not very subject to warp, and is very elastic and tough, as well as of, in general, a straight and even grain.

Chesnut is a wood scarcely inferior to oak for many purposes, being extremely durable, and not subject to decay, of generally an even grain and free from knots or curls, very tough, and of dimensions sufficient for most purposes of carpentry. Beams that have been taken from old buildings of more than a century standing, have been found as sound, and perhaps harder than when first placed in their original situation, and some posts that have formed a part of the same building, which have been always exposed to the different degrees of wet and dry, seemed little or nothing the worse, and, to appearance, as sound as when first the building was erected.

Beech is the last wood I shall here notice: it possesses qualities that recommend it for strength, evenness of grain, and toughness; it is not surpassed by any other English timber; but if exposed to the action of the atmosphere, it is not found to stand so well as many other woods; it is, however, particularly applicable to sweep work, and is on that account much used by millwrights, and might be used to many purposes of building with considerable advantage; it is also particularly adapted for pins in framing and doweling boards together, as it is not brittle or liable to snap, and drives well.

General Directions for Seasoning and Preserving Timber.

On the care we take in seasoning timber previous to applying it to the purposes of building, depends in a great measure the strength and durability of the structure we erect, for green or unseasoned timber must inevitably not only shrink and warp after your framing is put together, making the several joints loose, and straining the several parts of the framing, but also from due care not being taken in this particular, the dry rot is almost an inevitable consequence, bringing on premature decay in the structure; I would therefore recommend the following observations to the notice of the builder: first, that the timber should be felled at the proper season; secondly, that a sufficient time is given it before it is cut into the necessary scantling; and thirdly, that when thus cut it should be stacked so as the wind and air should have sufficient power over it, to dry up as far as possible the remaining moisture and sap that it contains; and on the first head, that timber which is felled in autumn is always the best, as at that time the sap is low, and consequently less of the juicy qualities are present, and the timber is consequently of a firmer texture, and not so subject to the dry rot; and though the objection to this season of the year (particularly with regard to oak) may have some weight, as the bark is not so easily separated, yet with regard to the art of building, though the timber may be somewhat enhanced in price, still we ought not to let the quality of it be deteriorated from this consideration, which, in my opinion, is but a secondary one, particularly as many suggestions have been brought forward, which, though not generally adopted, would be found not only advantageous to the timber itself, but the bark would be found superior in quality if means were adopted to bark the trees in the spring as they stand, and let them remain till the autumn to be felled; and I am confident some mechanical contrivance might be found to perform this operation with as great facility as the present mode in use, after the tree is felled.

With regard to the second head, the tree should remain at least till the following spring; but if longer the better, before it is sawn into scantling as timber, as time should be given for the action of the wind and sun to dry any moisture that remains from the sap contained in the pores of the wood. If possible, it should remain on the spot where it is felled; but, at any rate, no builder should cut his timber immediately it is drawn into his yard, unless previously laid some time in the field.

Lastly, in the third place, after the timber is cut into such planks and scantlings as it is wanted for, it will be necessary to let it dry further, and for which purpose we should expose it to such action of the air as will completely season it for use: this process should be, at the very least, six months; and as every builder has a different method of stacking, so it is necessary to select that which seems best adapted to the purpose intended. And first, with regard to planks, the general method is, as soon as cut, to nail a piece of wood at each end, to prevent them splitting, and then place them upon each other with a piece of pantile lath or other similar piece of wood between each, across the plank; or else place them side by side at a distance apart, so that the second tier or row, when laid on the first, shall rest only on the edges of the row beneath, and thus piled one above another, so that the air may have a free passage throughout the whole height of the stack, which is, perhaps, the best way, as the planks will not be so liable to warp in their length, though this method certainly takes up more room than the first. Boards are best seasoned by being placed on end resting between racks at the upper ends; and the same may be said with scantling of different dimensions, as all we have to do is to place them in such a situation that the air shall have free access to them, to dry up, as far as possible, the remaining sap or moisture they may have contracted; for if timber after being cut is suffered to lay close together, without admitting the free circulation of air, the effect will be, that the juices will cause a degree of heat to be evolved, which produces fermentation, and consequently a kind of premature decay, and which among workmen is called doaty timber, and which causes it to lose that firmness of texture which is natural to it, and become short or brittle; as, also, to be liable to decay much faster than timber properly seasoned. And here I shall take occasion to observe, that from the natural growth of timber, which consists of alternate layers of a spungy matter, and a harder substance which appears as rings arranged round the centre of the tree, that we must, where possible, consult the natural structure of timber to arrange it in the best possible manner to ensure strength and solidity: now these hard layers, or rings, are technically called the beat of the wood, and according as this is placed in our framing, so much the greater or less strength is obtained; a circumstance but little attended to, but which is at least worthy of consideration, both from mechanical and mathematical principles, as we might show, that in a beam where the beat is perpendicular to the horizon, the strength would be far greater than in one where it was in a parallel direction, or that it would be much less liable to bend in the former position than in the latter; for any rafter or beam whose sides are in the proportion—for instance, as three to one, will bear a greater weight when it is placed with its widest side perpendicular to the horizon, than when in a contrary position, so, in like manner, if we suppose it to be square, and on the supposition that the beat is the firmest part of the wood, if the beat is placed perpendicular, the beam or rafter will be much stronger than if it is placed parallel to the horizon, nearly in the same proportion as the beat is to the porous parts of the timber; the truth of this every workman may prove; by taking a shaving off the end of a plank and breaking it, he will find that it separates much more easily when bent in the direction of the beat than when in a contrary direction; this circumstance, though not generally noticed, is, I think, worthy some consideration, for the business of the carpenter is to combine with the least possible quantity of timber the greatest strength; and as a saving of material with requisite solidity is of material consequence, I wish to impress on the minds of the workman everything that will tend to improve the practice and application of his art.

Charring Timber.

This process is useful for such timbers as are partly placed in the ground, such as posts or timber used in the foundation of buildings, as the carbonic coat or charcoal covering which incrusts it by this process, preserves for many years what otherwise from continual moisture would sooner decay. The method of doing it most generally in use, is to apply actual fire to the part intended to be charred; by lighting a fire around it and letting the timber burn till its surface is sufficiently charred, and then either pouring water over it, or covering it from the air, it will be found to be burned a small way below the surface.

ANOTHER METHOD.

By giving a coat of pitch over the part intended to be charred, and setting a light to it, let it burn out, keeping it turned so that the fire shall have effect on all sides, and if necessary repeat the process, till it is sufficiently burned.

Note.—A coat of pitch is sometimes substituted in the place of charring, but that is far inferior in its effects: sometimes after the charring a coat of pitch is added, but that, in my opinion, is not necessary when it has been well charred.

A Composition for Weather Boarding Paling, &c.

Take of pitch six pounds, grease one pound, red ochre a sufficient quantity to colour it, and, if you like, a little lamp black, put it on hot, and when the boards are dry, otherwise it will not adhere.

ANOTHER.

Take of common tar eight pounds, fine sand two pounds, mix them well together, and colour the mixture by adding some ruddle or red lead and a portion of common soot.

A SUPERIOR COMPOSITION FOR THE SAME PURPOSE.

Take of common fish oil one gallon, add of lime a sufficient quantity to make it of a thickish consistence, to which add a sufficient quantity of yellow ochre and lamp black to make it of a greenish colour to your mind; this may be thinned to a proper consistence with linseed oil, to admit of being laid on with a brush, and will be found a very cheap and durable paint for the purpose of outdoor work.

A Composition for Preserving the Joints of Framing exposed to the Action of the Weather.

Take of pitch one pound, fat or grease a quarter of a pound, melt them together, and add finely powdered chalk to make it of a pretty stiff consistence; spread it over your tenons and within the mortices previous to pining your work together; it will thus prevent the ingress of moisture and preserve your framing for a considerable length of time.

ANOTHER METHOD.

Take one pound of white lead and mix with two pounds of glue, let them well boil and then strain; this is a more simple and much readier process, and for joints of sign boards, shutters, entablatures, and all exterior work, is by far the best.

To Render Wood Incombustible.

A very excellent way to render wood incombustible, is to soak it in a strong solution of alum and the sulphate of copper. About one pound of alum and one of the sulphate of copper should be sufficient for 100 gallons of water. These substances are dissolved in a small quantity of hot water, then mixed with the water in the vessel in which the wood is to be steeped. The timber to be rendered fire-proof can be kept under the liquor by stones, or any other mode of sinking it. All that is required is a water-tight vessel, of sufficient dimensions to hold enough of liquor to cover the timber, which should be allowed to steep for about four or five days. After this, it is taken out, and suffered to dry thoroughly before being used.

To Preserve Iron Work, as Bolts, Straps, &c.

To two quarts of linseed oil add half a pound of litharge, let it simmer over a slow fire for two or three hours; take it off and strain it, then add of finely powdered rosin a quarter of a pound, and white lead one pound, keep it in a gentle warmth, stirring it frequently till the rosin is dissolved and the whole well incorporated; then, if with this mixture you smear the bolt or strap previous to placing it in its place, and put a good coat on after on the parts exposed, your iron work will be effectually preserved from rusting from the effects of the atmosphere, or any wet that may get to your work.

PART II.—JOINERY.

THE business of the joiner is distinct from that of a carpenter, insomuch as it regards the more ornamental and nicer parts of the art of building; whereas the carpenter is concerned more with the solidity and stability than the beauty or decoration necessary to a finished piece of building. I shall, therefore, endeavour to be minute in the instructions, and particular in the directions, that accuracy and necessary solidity may be combined, that beauty and truth of workmanship may be attained; and, as we here mean not to dictate to the experienced workman, but to throw out hints and produce examples to assist those who may not be acquainted with the various methods in use, no apology will, we trust, be necessary for introducing what may be generally known amongst the more initiated, but for the benefit of the less experienced endeavour to compress in a small compass all the material rules and directions that tend to produce a good workman.

The different kinds of wood mostly in use with the joiner are white and yellow deal, wainscot, or American oak, and mahogany, which, though unnecessary to describe particularly, I will here mention their qualities, and uses to which they are applied; and first, with respect to *deal*, which is generally imported into this country in lengths of from six to fourteen feet, though twelve feet is the most general length, and for most purposes cut to the greatest advantage; the thickness is about three inches, and width about nine. In choosing deals, we must select those free from knots or shakes, and which appear of the straightest grain, reserving the coarser ones for such common purposes that we may have occasion for. The yellow deals are in general the straightest grain and freest from knots, and consequently most proper for styles of doors and framing, as well as for sticking all sorts of mouldings, as it works clean and sharp. The white deal is proper for panels, which should be cut down the middle and joined again; an edge to a middle and so on, to the width of the panel, and carefully avoid a knot, if possible on the edge of your joint, as it will frequently cause it to fly, however good it may be at first, if the panel should shrink ever so little. The next wood in order is wainscot, which is imported in logs of different dimensions; such should be selected as appear to be cut from a straight tree, and which should be of a clear grain, and free from streaks of a lighter colour, as then parts are of a softer texture, and are the beginning of decay in the timber, termed among workmen doughty parts of the wood. In cutting a log of wainscot, if we cut it in the same direction as the beat of the wood, the boards will be variegated and have a handsome appearance; if we cut it contrary to the direction of the beat, it will have an uniform appearance and a clear look; the first is proper for panels and such like, and the latter for the styles and frames of doors, sashes, &c. as it will mortice better, and be less likely to split or warp. The last wood we shall mention is mahogsny, which is used only in the best kind of work for doors, sash frames, banister rails, &c., and is of several kiuds, known principally by workmen under the general terms *Spanish* and *Honduras*. The first is considerably harder than the last, and is less frequently of a variegated grain, though when it is, it is very handsome. The latter is generally of a spongy texture, and and often very cross grained, which contributes to its mottled or variegated appearance, and is often very difficult to work to bring it to a smooth face, but when good is very proper for panels of doors, &c., though the Spanish is by far the best for framing or for mouldings, as it works clean, and is generally used for the banister rails of staircases, as it is less liable to break when cut out on the sweep, as its texture is both strong and its grain even. There is another kind of mahogany, known under the term *Ratteen*, which is often of great use for panels, as its dimensions are large, therefore there is no occasion for jointing, and as it is of a reasonable price, is often substituted for deal, particularly for sweep work that is to be painted, as the facia of shop fronts, sign boards, &c., that are liable to the changes of wet and dry, or are exposed to the sun and air; for however good your joints may be, they will, in this case. be liable to fly, however well secured by blocks at the back of the joint, or other means generally used. Having given a short sketch of what regards the woods in general used by the joiner, as well as the uses to which the different sorts are most applicable, we will now proceed to give some useful receipts applicable to this part of the building art.

Polish for Wainscot Work.

Take of the best yellow bees-wax, shave it with a plane into thin slices, and put it into a glazed earthen pan. add as much spirits of turpentine as will cover it, and let it dissolve without heat, stirring it occasionally, it will then be of the consistence of butter; if too thick, add more spirits, if too thin, more wax; this mixture must be applied with a linen rag on the places you wish to polish, rubbing it well into the grain; after which, with clean cloths, either linen or woollen, rub it well till it has a good gloss and does not feel sticky if the finger is applied to it.

Oil for Wainscot Work.

Take of the best linseed oil one quart, to which add half a pint of the best spirits of turpentine, and a piece of lime about the size of a cricket-ball, broke in pieces; let it simmer on a stove or near the fire, covered over for two or three hours, then strain it through a coarse cloth and keep it for use.

It must be put on with a brush, and when it has lain on the work about twenty-four hours rub it off with a woollen cloth, and some oak saw-dust, and finish with a clean linen rag.

Another Oil to Heighten the Colour.

Linseed oil one quart; litharge half an ounce; let them simmer together for an hour or two; then strain it off; take now about half-a-pint of spirits of turpentine, and put to it as much as you will of pounded turmeric, till a sufficiency of the colour is extracted: strain it off, and add it to the oil; it will be found excellent, as it heightens the colour of the wood, and is to be used the same as the foregoing.

N. B.—The process may be repeated two or three times, letting a day or two intervene; it will materially add to the beauty of the work.

Polishing Oil for Mahogany.

Take linseed oil, one quart; alkanet root, one ounce; rose pink, half an ounce; stir them well together, and place them near the fire to simmer gently for an hour or two; then strain off in a clean pan. Apply with a brush and let it remain for about an hour; then take of the finest red brick-dust sifted through a cloth or sieve, and dust it over your work; polish it with a piece of woollen cloth by rubbing it well the straight way of the grain: after, finish with saw-dust and a clean cloth.

N.B.—Some prefer polishing mahogany with bees-wax and turpentine, as directed for wainscot.

To Clean up Deal Panels, &c.

After your work is planed as smooth as possible, apply hot size, or very thin glue, and let it dry; then with a piece of hearth-stone rub it well, and you will produce a very smooth face, which will make the painting, afterwards to be applied, appear smooth and even; this is much better, as well as a more ready way, than that of sand or glass paper.

To make Glass or Sand Paper.

As the paper for the purpose of cleaning off work, known by this title, is of great use to the joiner, we will here give the process of manufacturing it, as it is seldom to be met with very good. Take any quantity of broken glass, that with a greenish hue is the best, and pound it fine in an iron mortar; have ready three sieves of different degrees of fineness; take several sheets of paper, free from knobs, fine cartridge is the best, and brush them evenly over with thinnish glue, then either hold them to the fire or lay them on a hot piece of wood, and sift the pounded glass freely over them through the finest sieve of pounded glass; let them remain till the glue is set, and shake the superfluous powder off, which will do again, and so you may proceed with the other sieves for different degrees of fineness; hang them up to dry and harden, and you will have a superior kind of paper to that in general use, as the pounded glass is often mixed with sand, which greatly injures the quality of the paper, and produces scratches when used for cleaning off your work.

Stone Paper.

As in cleaning of work made of deal, or soft wood, sometimes one process is found to answer better than another, the following will be found often very useful, as it makes a kind of sand paper, which in some cases will answer very well, as it is very fine, and at the same time produces a good face on the wood to which it is applied. Having prepared your paper as directed in the last receipt with glue, take any quantity of pumice stone, and having pounded it, sift it over the paper, through a sieve of moderate fineness, then let it harden, and repeat the process till you get a tolerably thick coat on the paper, which when perfectly dry, will be found to be a very superior paper for polishing your work, as it is not liable to leave scratches, but leaves a smooth and even surface.

Glue.

The quality of glue being of material consequence to the joiner, it may not be amiss in this place to say something respecting it, and the tests by which we may ascertain its adhesive properties, as by this means we may be enabled to select that which is best, as well as to reject that which does not possess the requisite qualities of adhesion and firmness, and first it may be observed that glue is made from either the skins or sinewy parts of animals, and also from the skins and some other parts of fishes, that of the eel and shark; that from animal substances is reckoned better than that from fishes; though the strongest glue perhaps, we are acquainted with, is isinglass, which is made from the air bladders of a species of large fish found in the Russian seas, but its great price makes it of little use to the joiner when other glue can be substituted; however, from chemical experiments that have been made, that glue which is manufactured from the skins of animals, is superior to that which is made from the other sinewy or horny parts of animals, and which is found by actual observation, in practice to be much superior to the glue made from the skins, &c. of fishes, as it is not so subject to be affected by the moisture of the atmosphere; therefore the workman will always prefer animal glue to what is generally termed fish glue, but which latter is often sold as glue of the best quality; and here we shall endeavour, first to lay down some directions to choose this necessary cement for the joiner, and give such directions as shall enable the workman to form some estimate of its adhesive qualities; all glue in the cake is subject to the effect of the dryness or moisture of the atmosphere, becoming soft in damp weather, and crisp in dry; but the different kinds are differently affected, therefore it is best to purchase in dry weather, as that which is then soft is not of such good quality as that which is crisp, and if we hold a piece of glue up to the light, that which is the most transparent is in general the best; and here it would be advisable before making a purchase, to submit to experiment a sample of the article which you wish to purchase ; thus, if we take a cake of glue and cover it with water in a pan, and let it remain for two or three days, if it is good, it will not dissolve at all, but will swell by being laid in water; whereas, that which is of inferior quality, will partly if not wholly dissolve in the water, for that which least dissolves is the best, or possesses superior qualities of adhesion, and will be least affected by damp or moisture; another test is, that being dissolved by means of heat in water, that glue is the best which seems most cohesive, or which is capable of being drawn out in thin filaments, and does not drop from the glue brush as water or oil would, but rather extends itself in threads when falling from the brush or stick, which, if the glue possesses the requisite properties, will be found to be always the case; these few hints, with a little experience, will enable the workman to judge of the quality, as well as the method of selecting that which is best calculated to ensure success with regard to the firmness and stability of our work. We may here add, that that glue which is made of the skins of old animals, is much stronger than that of young ones.

Glue to hold against Fire or Water.

Mix a handful of quick lime in four ounces of linseed oil, boil them to a good thickness, then spread it on tin plates in the shade, and it will become exceedingly hard but may be easily dissolved over the fire, as glue.

To make a very strong Glue.

Take an ounce of the best isinglass, dissolve it by moderate heat in a pint of water, strain it through a piece of cloth, then add of the best glue in cake, which has been previously soaked for twenty-four hours, and also a gill of the strongest vinegar; let the whole dissolve by placing it near the fire; after it is dissolved let it boil once up, and strain off all the impurities; this will make a glue which may be reserved for that part of your work which requires particular strength, or where the joints themselves do not contribute to hold the work together, such as small fillets and mouldings, or carved patterns that are merely held on the surface by the glue.

To Glue Joints.

In general nothing more is necessary, after ascertaining your joint is perfectly straight, and as is technically called, out of winding, than to glue both edges, with the glue quite hot, and rub them lengthways till the glue is nearly set, but not chilled; however, when your wood is spungy, or sucks up the glue, the following method will be advisable, as it strengthens the joint, and does away with the necessity of using the glue too thick, which should always be avoided, as the less glue there is between the joint, provided they touch one another, the better, and when the glue is thick, it sooner chills, and we cannot well rub it out from between the joints; the method is to rub with a piece of soft chalk each joint on the edge, and wipe it off again with your finger, so that no lumps remain, and then glue it in the common way; it will be found to hold much faster, particularly when the wood is porous, than when glued without the chalk.

To make Cement for Stopping Holes and Flaws in Wainscot, &c.

Take of bees-wax and pounded resin equal parts, dissolve them in a pipkin, gently letting them incorporate, and stir them with a stick till intimately mixed and dissolved; then for wainscot add chalk, yellow ochre, and umber in powder, till the colour is to your mind; for mahogany, add instead of the umber a little red ochre, and not so much chalk, or a little burnt umber will sometimes make it nearer the colour you wish.

Another Cement much better.

Take of fine saw-dust of the same wood you desire the cement to imitate, let it macerate or soak in water for two or three days, then pour some of the water off, and place it in a pipkin on the fire covered over, and let it boil till it becomes quite a pulp, strain it through a cloth, and press as much of the water from it as possible, and keep it for use, and when wanted, mix it with hot glue to a proper consistence, and fill the cracks in your work, which if properly applied and left to get quite hard, will scarcely be distinguished from the wood itself.

Of the different Methods of Joining Timber.

Every workman must be aware of the meaning of the terms dove-tail, mortice and tenon, grooving, &c., but the best methods of performing these several operations of joining their work together, they only get by experience, and are not in general aware of the proportions that one piece should be which is fitted into another, so as to produce the greatest strength with the least waste of material, or so to proportion their joints, that one part shall not be liable to fail or give way before another: we shall here therefore endeavour to lay down some rules, and produce some examples that will be an attempt, at least, to bring into view the principles of the mechanism of joining, the absence of which is often the cause of work not standing well, and cause the parts either to separate with a trifling strain, or from being bound too tight together, to fly and split in all directions, not so much in general from the bad execution of the work themselves, as from want of proportioning the strength to the stress of the joints; we shall therefore arrange in order the several kinds of joints or methods of framing and joining timber, and under each head give such directions, founded on the principles of mechanics, as will enable the workman to proceed with some degree of certainty, and not, as is too frequently the case with many artizans, observe no rule but that which custom has authorised, or practice made familiar.

Dove-tailing.

As Dove-tailing is of great use in the art of joinery, I have represented in Plate I. several sorts; now as much depends on the proper proportioning the parts which fit into each other, so that the pin or socket (that is, the part represented in fig. 1, called the pin of the dove-tail, and that in fig. 2, called the socket,) shall be as nearly as can be of equal strength. I shall lay down some rules for the guidance of the workman, and shall here refer to the pin only at fig. 1, (for the socket is made to correspond to it); let A B C D be a scantling, which is required to be joined to another by means of a single dove-tail; now as much depends on the form of the dove-tail, as well as the proportion it bears to the parts cut away, I shall endeavour to lay down the principle on which the greatest strength is maintained; having squared the end of the scantling, and gauged it to the required thickness A I K L M, divide I M into three equal parts at K L; let K L be the small end of the dove-tail, and make the angles I K G and M L H equal about 75 or 80 degrees, and make G E and F H parallel to A N and B O, where enter the saw, and cut away the pieces A I K G E N and B M L H F O, and having cut fig. 2 to correspond by marking the form of the dove-tail on the top of the piece A B C D, it will fit together as shown in fig. 3. We may here observe, that according to the texture of the wood, we may make the bevil of the dove-tail or angle I K G, fig. 1, either more or less than I have mentioned. Hard close-grained wood, and not apt to rive or split, will admit of a greater bevil than that which is soft or subject to chip; thus the dove-tail in deal must be less beviling than that in hard oak, or mahogany. And it is a great fault in many workmen in making their dove-tails too beviling, which instead of holding the joint firmer together, weakens it; for provided the bevil is such that will prevent a possibility of pulling the pieces apart, in general we may observe that the less bevil is given the better, and this may be observed if we compare the dove-tailing of the cabinet-maker and the joiner; the former has very little bevil, while the latter is very much so; and also with regard to the appearance of the work itself, the one looks neat (and is at the same time strong,) while the other appears to aim at great strength, though at the same time looks clumsy and is much the weaker.

Fig. 4 represents the dove-tail in common use for drawer fronts, &c.; where we wish to hide the appearance of the joint in front, the board A B C D is cut with the pin, and A E F B, with the socket; the pins in this sort of dove-tail are in general from about three-quarters of an inch to an inch apart, according to the magnitude of the work in hand. Fig. 5 represents the pin part of a *lap* dove-tail, which when put together shows only a joint as if the pieces were rebated together as shown at fig. 6; the part A B C D represents the pin, and the part E F G H the socket dove-tail, and when put together only shows the line H G, as a joint; and if the corner A B is rounded to the joint G H, it will appear as if only mitred together; this kind of dove-tail is very useful for many purposes where neatness is required, such as chests, boxes, &c. Fig. 7 is a still neater dove-tail, and as the edges are mitred together, is termed a *mitred dove-tail;* it is the same nearly as the last figure, only that instead of the square shoulder or rebait in A B, it is cut into a mitre and the other piece is made to correspond.

Another very neat, as well as expeditious method of joining pieces, and which is somewhat analogous to dove-tailing, is shown at fig. 8, where the joint is first formed into a simple mitre, and then keyed together, either by making a saw kerf in a slanting direction, as shown at A B, or by cutting out a piece as at C D, in the form of a dove-tail, and fitting a slip in that, of the required form; the first method, as A B, is amongst workmen called keying together; the second, as C D, is key dove-tailing.

The last method I shall describe is shown at fig. 9, and may be termed *mitre dove-tail grooving,* the part A B being formed with shoulders cut to the required bevil, and a piece left for the pin dove-tail, which is inserted into the socket dove-tail made to correspond to it in the piece C D, which has been previously formed into a mitre; this method, though not much employed may be used with great advantage in many cases, particularly when we wish to join any pieces together the lengthway of the grain.

Mortice and Tenonding.

Under this head I shall endeavour to lay down some rules to be observed with regard to proportioning the parts of the mortice and tenon, so that they shall both be equally strong, or that the tenon shall not be more likely to give way than the cheeks of the mortice, for on this principle depends in a great measure the soundness of the work; and as what is laid down is formed from actual experience and practice, it will, if not always found quite correct, be in general a safe guide for the workman, and will prevent that frequent error of allowing too little substance for the tenon, for fear of weakening the cheeks of the mortice; and I may here observe that this subject is well worthy the attention of the ingenious mechanic, as well as the consideration of the mathematician, but as the latter is foreign to the purpose in hand, I shall proceed with the practical part, not without the hope that it will induce the workman to be more particular in general with regard to the proper proportioning the several parts of his work, so as to approach as near as possible to a maximum of strength with a given quantity of material.

Figs. 1 and 2 represent a simple *mortice* and *tenon;* the dotted lines show the parts to be cut away; now to show the thickness of the tenon and consequently the width of the mortice, we have here one tenon and two shoulders, which is three parts, one part of which is to be allowed for a tenon and two for the shoulders; and this in general will be found the best proportion, for if the tenon is more than that, it will weaken the shoulders of the mortice, and if less, the tenon itself will be diminished in strength, and will be liable to break off, with a force that would not split or separate the shoulders of the mortice; now if we have, as is frequently the case, two tenons in one piece, as shown at fig. 3, as there are two tenons and three shoulders, which is five parts, which shows that each tenon must be one-fifth of the thickness of the stuff, and the shoulders are all equal to the tenons, and this rule may be generally observed, unless the tenon is at a considerable distance from the end of the stuff, and then something more may be allowed for the thickness of it, as the mortice in that case is not so liable to split; but it should in no case, however sound the timber or tough the material, be more than two out of four parts, that is, it would never be safe to make the tenon more than half the thickness of the stuff, and that only under particular circumstances, and when the mortice is near the middle of the scantling, or we should considerably weaken the piece in which the mortice is cut.

There is frequently in joiners' work, a shoulder at the *bottom* of the tenon, which fits into the piece in which the mortice is cut, as shown at fig. 4, and the tenon is divided into two parts as there shown, which, when the stuff is wide is a good method, as it strengthens the piece in which the mortice is cut without weakening, in the same proportion as the mortice itself; and I would advise in this case, that the piece **B C**, cut out from between the tenons

A B and B C, to be nearly, if not quite, one-third of the distance A D, as if much less, the piece left between the mortices will add but very little to the strength of the piece in which the mortice is made; and hence we should have the tenon stronger than necessary, in proportion to the morticed piece; and here I may observe, that if the width of the tenon is much more than four times its thickness, we shall gain additional strength from dividing the tenons into two or more parts, as shown in the figure, particularly if we allow a small piece at the bottom of the tenon as represented in this figure.

Grooving and Lapping.

It is scarcely necessary in this work to say much on this part of joining timber or boards, it being analogous to that of morticing and tenonding. I shall therefore under this head, simply state, that when we wish to join two boards together, by means of a tongue and groove, the groove should never exceed one-third of the thickness, and often if the piece which is formed for the tongue is hard wood and not liable to split, one quarter of the thickness will be sufficient: or in the case of a panel let into a groove in the style, we must be often guided by the thickness of the panel itself, which should never be less than one-third the thickness of the style.

In making a groove across the grain for partitions, &c., it will be best in most cases to make the groove about a fifth or sixth of the substance of the stuff; but if the groove is formed into a dove-tail, one quarter the thickness will be best, and the dove-tail should be made a very small degree tapering, but not too much, and only so as to go almost home without requiring a blow from the hammer or mallet, to drive it into its place till very nearly so, as all joints should only fit so tight, that before they are glued, they can be easily separated with a gentle blow. With regard also to a lap joint, or lapping two pieces together, supposing them of equal thickness, half the substance of each should be cut away; and if of unequal thickness, we should make the lap in the thinner piece, about two-thirds or three-quarters of its thickness, according to the substance of the thicker piece, thus endeavouring in this as in all other cases, to avoid weakening one piece more than another.

Bending and Glueing up.

With respect to bending or glueing up stuff for sweep work, much judgment is necessary; and as the methods are various, I shall mention a few, that the workman may apply them as occasion requires, one method being preferable to another according to the nature of the work in hand.

The first and most simple method is, that of sawing kerfs or notches on one side of the board, thereby giving it liberty to bend in that direction; but this method though very ready and useful for many purposes, is still very weak where any strain may be on the piece. Still, in this instance, we may in some measure make a tolerably strong sweep, if after sawing the kerfs, and being particular to make them regular and even, and sawing them at equal depths, we rub some strong glue into each kerf, then bend it to the required sweep and glue a piece of strong canvas over the kerfs themselves, leaving the glue to harden in the position which we have bent our stuff to.

The next method is, that of glueing up our stuff in thin thicknesses, in a cawl or mould made with two pieces of thick wood cut into the required sweep; and this method, if done with care—that is, making the several pieces of equal thickness throughout, and free from knots is perhaps, the best that can be devised for strength and accuracy. It is also a practice sometimes to glue up a sweep in three thicknesses, making the middle piece the contrary way of the grain to the outside and inside pieces, which run lengthways. This method, though frequently used for expedition, is much inferior to the above, as it does not allow the different pieces to shrink together, and consequently the joint between them is apt to give way. Again, in many instances, a solid piece, if not too thick, may be bent into the form required; if we soak well the outside of the curve with hot water, and hold the inside to the fire, when having formed the curve to your mind, you retain it in that position till cold and dry, it will retain the curvature given to it.

The last method I shall here mention is, that of forming a curve by means of cutting out solid pieces to the required sweep, and glueing them upon one another till you have attained the thickness required, taking care the joints are alternately in the centre of each piece below it, something in the manner of a course of bricks above each other; in this case it will be necessary, if the work is not to be painted, to veneer the whole with a thin piece after the first has been thoroughly dry and planed level, and also made somewhat rough with either a rasp or toothing plane

Scribing.

By scribing is meant, generally, the method of making

one piece of stuff fit against another when the joint is irregular; thus the plinth of a room is made to meet or correspond with the unevenness of the floor; in this manner, by opening your compasses to the greatest distance the plinth is from the floor where some parts touch it, and letting one leg run along upon the floor or uneven surface, the other leg will leave a mark on the plinth, which if we cut away the stuff to that mark, it will then make a good joint with the floor; but the great use of scribing to the joiner is, that of joining moulding of panels or cornices that shall, when placed together, seem a regular mitre joint; and it has this advantage over the common method of mitreing—that if the stuff should shrink, it will scarcely alter the appearance of it, whilst that of the mitre, under the same circumstances, causes a gap to show itself, and the joint to appear bad. The method is this: to cut one piece of the moulding to the required mitre, and then, instead of cutting the other to correspond to it, cut away the parts of the first piece, till we come to the edge of the moulding, which will then fit as the other moulding, and appear as a regular mitre.

Finishing of Joiner's Work.

As much depends on this part of the operation of joining, I shall give some hints to the workman, that for want of paying proper attention to, however well the work may be executed with regard to its strength or the accuracy of the several joints, will appear but slovenly executed, if the finishing is not well attended to, whether it is intended to remain of the natural appearance of the wood, or afterwards subject to the process of painting or varnishing; and first, with regard to those pieces of work composed of wainscot, oak, or mahogany, here our chief aim is to make the surface perfectly smooth and even; now, in order to avoid a deal of trouble in this part, it will be found necessary, after having glued your framing, &c. together, to let the glue that ooses out and is spilt about, first remain a few minutes to chill, and then you can carefully scrape it off with a chisel; and with a sponge dipped in hot water and squeezed nearly dry, clean out all the quicks and corners that cannot be got at with the chisel; this will not only save a deal of trouble in the after operations, but will prevent a stain that often is left if the glue is suffered to remain till quite hard; particularly on wainscot, which will turn black in every joint or place where the glue has been suffered to remain. After this operation, which, though it may appear tedious to some workmen, will be found in the end a saving of time, let your work remain till perfectly dry, having levelled your joints and other parts with a fine smoothing plane, scrape the whole surface with a smooth scraper, and finish with fine glass paper. It will be sometimes necessary, where the grain is particularly cross, as in some mahogany, to damp the whole over with a sponge to raise the grain, and then again apply the glass paper. Your work will now be ready for polishing with wax, oiling, or varnishing, and according to the pains taken in this part, will the work appear.

Now, with regard to cleaning up deal or fir, the same precautions may be taken with regard to clearing off the glue, and the other parts may be then smoothed with a piece of glass paper that has been rubbed with a piece of chalk, or some workmen prefer, for many parts, to rub with a piece of hearth-stone; it will then be ready for the painter; but as there are many knots and places where the turpentine would ooze out and spoil the appearance of the paint, those parts are done over with a composition which is called priming; this, though properly the painter's business, is often necessary to be attended to by the joiner. The composition in general use for that purpose is made with red lead, size, and a little turpentine; to which is sometimes added, and is an improvement, a small quantity of linseed oil; this prevents the knots showing through the paint. Some workmen omit the oil and turpentine, but that is bad, as the size by itself will be apt to peel off, and not insinuate itself into the parts.

Another good method of cleaning off deal is, after having made the surface quite smooth with the plane, to rub it with a piece of chalk, and with a fine piece of pumice stone to clean the whole as with glass paper, and if the grain should still appear rough, damp the whole with a sponge, let it dry, and repeat the operation.

I have been particular in this place to impress on the workmen the necessity of cleaning up their work well, as the present taste for internal decorations is that of imitating different fine woods and marble, which will not look well unless particular care is taken in making a good surface for the artist to lay his colours on, as every defect in the ground will show itself through them more than in the common way of using a body in the colour, and giving several coats; but even in that case, the work that is well prepared will not only look better, but the colour will not be so apt to chip and peel off, as when the surface is not properly levelled.

PART III.—CABINET MAKING.

Much that has been said on Joinery applies also to Cabinet-making as respects mitreing, dove-tailing, &c. &c. The general term Cabinet-making is the art of making all such parts of the furniture of a dwelling-house as are made of wood, together with the art of Chair-making, &c., and in order to arrive at any degree of perfection, the knowledge of designing,—carving,—modelling, &c., is requisite.

It has also been supposed that a knowledge of geometry, and particularly of that portion of it which treats of the description of curved lines, is of great use to the cabinet-maker; but, with the exception of a knowledge of perspective, and of a few simple methods of drawing common curves, geometry is of little use to him; and, when it is studied too closely, it leads to a harsh and mechanical mode of designing.

The best advice we can give the cabinet-maker, in acquiring a graceful, easy, and free method of drawing, is, to draw as much from nature, or from good casts, as possible. It is not of material consequence whether vegetable or animal forms be drawn, but a mixture of both is desirable, as they have very distinct characters, which will be easily traced in attempting to delineate them

General Remarks on Designs for Cabinet Work.

In Design, the central or principal part of the object requires most notice. The other parts should be so fai subordinate to it as not to distract the attention from the centre; and, yet they should be so united in harmony with it, as to be obviously essential to complete the design.

The connection between the principal and the inferior portions of the design should be preserved by the continuance of some of the leading lines of the principal part to the inferior ones; and, whether these lines be straight or curved, they should never be so far interrupted by ornament as to render it doubtful whether or not they are continued; and, as the idea of firmness or stability is a necessary accompaniment of good taste in the design of furniture, the leading lines of the principal part of the design should descend in such a manner to the base as to give an idea of firmness, as far as the nature of the thing requires it.

Proportion, as it depends on the relative *magnitude of parts*, is, sometimes, wholly left to the good taste of the designer; and, when cases occur where it is within his power, one part in a design must form the principal object, and ought not to have a rival in magnitude; also, when the piece of furniture is seen in its best position, this principal part should be as near the centre of the whole as possible.

The principal part of a design should be sufficiently prominent for the eye to pass from it to the whole, or the reverse, without perceiving the change of magnitude to be abrupt; and the same remark applies to the relation of the subordinate parts of the design to the principal one.

If this attention be given to the proportion of the parts so that the eye may pass from the consideration of one to another, and not feel the change abrupt, the design will be pleasing.

If too small a proportion be assigned to the principal part, the design will be flat and unmeaning. If the proportion be too large, the whole will be absorbed in the part, as a modern mansion is not unfrequently all portico. A due proportion of the principal part to the whole given boldness and propriety.

Richness is produced by introducing as much ornament as the object will bear, without destroying the relation between the plain and ornamental parts; a design, overcharged with ornament, becomes frittered, and wants both variety and repose.

The opposite quality to richness is meagreness, or a deficiency of ornament; and want of attention to its proportions. Between the extremes of overcharging and meagreness, an immense variety of degrees of combination of ornamented with plain surfaces may be selected.

When the ornament consists of moulded work only the piece of furniture is termed plain; but, in rich furniture, the combined effect of moulded and carved work is necessary. In either species, the proportions of the ornamental and plain parts to each other should be regulated by like principles as the magnitude of the parts.

Coloured Woods, Metals, &c.

Sometimes richness of effect is no further attempted than is obtained by the natural beauty of the wood which is employed; and when this natural beauty is considerable, this simple kind of furniture is most highly valued.

But wood, so fine in colour and figure, as alone to give richness of effect to furniture, is very rare, and still more frequently defective; hence, the more usual mode of combining different coloured woods, or of metals and shells with woods, require some degree of attention. The prevailing combinations are formed by coloured bands, lines, and ornaments of wood, or by lines, beads, or ornaments of brass; the brass being in many instances cut into beautiful forms and further embellished by engraved lines on its surface.

The circumstances to be attended to in forming these combinations, are, harmony of colour, due proportion of the coloured parts to one another, and relief by contrast.

Much depends on the colour of the principal mass of the piece of work, which we call the predominating colour. If this colour be rich, very little variety of other colours should be added. On the contrary, if the predominating colour be light and delicate, it will bear to be enlivened and supported by contrast with fine lines or borders of an opposing colour; taking care that the mass of opposing colours be so small as not to produce opposition instead of contrast; for contrast, skilfully managed, gives force and lustre to the ground, while opposition destroys even its natural beauty

Framing.

Framing, in cabinet-making, requires the same precautions as in Joinery, when it is required to form large surfaces, for, owing to shrinkage, and warping of wood, large even surfaces can be formed only by means of pannelling.

The width of the style of a frame should be one-sixth of the whole width of a compartment of the frame; the tenons should be one-fourth of the thickness of the framing, and the width of a tenon not more than five times its thickness.

But, where surfaces of considerable width are to be formed without an appearance of framing, whether those surfaces are to be veneered or not, we should avoid framing them with other pieces where the grain of the wood is in the contrary direction, for the difference of the shrinkage of the two ways of the wood is so considerable, that it can scarcely be expected to stand without either warping or splitting when confined. Where warping is to be prevented, we strongly recommend that holes should be bored through, and strong iron wires inserted, at short distances apart, across the piece. These would act as clamps in preventing warping, and, at the same time, would not be affected by the shrinkage in width.

Angles are formed in various ways, depending chiefly on the object of the work. External angles of mouldings are either simply mitred, or rebated, or both rebated and mitred together. Internal angles are generally grooved together, with the outer edges mitred. Where the front edge only is to be mitred, a dovetail groove is made, and rather narrower at the back than at the front, so that the tongue tightens as it is driven in.

When a strong firm connection is wanted, and the wood is to be joined end to end, dovetailing is to be preferred. When the dovetails are not to appear, they may be formed by the method called lap-dovetailing; and, when the dovetails are cut through, it becomes the kind used to join the angle between the front and end of a drawer. When a joint is to appear as if it were mitred, the method of dovetailing employed is called mitre-dovetailing. The apparent edges are in this case always mitred to a depth of about an eighth of an inch. There is also the method of joining by keys; the parts being neatly mitred, then saw-kerfs are to be made for the slips of wood called keys, which are to be inserted with glue when the joint is put together.

Drawers are mostly dove-tailed together, but variously made in other respects. Well-seasoned wood should always be used, as otherwise the drawers are liable to break at the joints; the tenons should always be in the direction of the grain of the wood. In morticing, care must be taken that the mortice and tenon are neatly fitted, neither too loose or too tight, and the parts well glued when put together.

Veneering, Banding, &c.

Veneering is the method of covering an inferior wood with a surface of a very superior kind, so that the parts of the article of furniture thus manufactured which meet the eye, appear to the same advantage as if the whole work were of the best description. If this be well performed, it is very durable, looks well to the last, and is attainable at an expense considerably less than a similar article would cost if manufactured of the same wood throughout, but of an inferior quality.

The principal requisite to ensure success in veneering, is to select well-seasoned wood for the ground, and to use the best and strongest glue.

Veneers are worked either by a veneering hammer or by cauls. In veneering by the hammer, the ground should be warmed by the fire, and the outside of the veneer wetted with warm water or thin glue, with a sponge, and the side to be laid covered with a coat of thin glue and warmed at the fire, the veneer should be quickly laid on the ground and worked with the hammer, backwards and forwards, till neither air or glue will come out Veneering with the hammer is preferable when the veneers are straight and even, but as that is seldom the case, work is generally done with a caul.

A caul is made of solid wood, shaped to the surface to be veneered; it should be well heated, then oiled and greased, it is screwed down upon the veneer, and the heat and pressure sends out the glue, causing the veneer to bed close to the ground. The veneers should be of an even thickness when worked by a caul, otherwise the glue will collect, and the work is liable to blister, it should not dry too quickly.

To Raise Old Veneers.

In repairing old cabinets, and other furniture, workmen are sometimes at a loss to know how to get rid of those blisters which appear on the surface, in consequence of the glue under the veneer failing or causing the veneer to separate from the ground in patches; and these blisters are frequently so situated, that, without separating the whole veneer from the ground, it is impossible to introduce any glue between them to relay it; the great difficulty in this case is to separate the veneer from the ground without injuring it, as it adheres in many places too fast to separate without breaking it. We will here, therefore, show how this operation may be performed without difficulty, and the veneer preserved perfectly whole and uninjured, ready for relaying as a new piece. First wash the surface with boiling water, and with a coarse cloth remove dirt or grease; then place it before the fire, or heat it with a caul; oil its surface with common linseed oil, place it again to the fire, and the heat will make the oil penetrate quite through the veneer and soften the glue underneath; then whilst hot raise the edge gently with a chisel, and it will separate completely from the ground: be careful not to use too great force, or you will spoil your work; again, if it should get cold during the operation, apply more oil, and heat it again: repeat this process till you have entirely separated the veneer; then wash off the old glue, and proceed to lay it again as a new veneer.

Banding is a term applied to a narrow strip of veneer used as a border, or part of a border, either to a large veneer, or to solid wood; in the latter case, a rebate is sunk for the banding. Banding is of three kinds: it is called straight-banding when the wood is cut lengthwise of the grain; cross-banding when the wood is cut across the grain; and feather-banding when cut at an angle between the two.

Between the banding and the central part, one or more lines are generally inserted, and sometimes a narrower band.

The joints of banding should be as well matched as possible, both in respect to colour and grain; and, excepting the mitre-joints, it is an advantage to make the joints at the veins of the wood.

Inlaying, &c. &c.

Inlaying is an expensive method of ornamenting furniture with fancy woods, metals, shells, &c., and if not well executed is unsightly and liable to frequent breakage. It is of great antiquity, and was brought to great perfection about the 16th century; it was revived about the end of the 17th century, in France, but met with little encouragement, though practised by some eminent artists

—amongst the most famous for the excellence and extent of his works was one Boalle, or Buhl, from whence we take the name of Buhl-work. It has been much in use in England during the last twenty years, to form ornamental borders, chess tables, &c.

In this art the part for the ornament, and that for the ground, are glued together, and the design being drawn upon one, both are at once cut through by a very fine species of bow-saw. Thus, there are four parts obtained, which, being put together in two, the one is the ornament designed in its proper ground; and the remainder of the ground, combined with the remainder of the ornament, gives another pattern called the reverse.

The plates of brass or other metal should be of the usual thickness of a veneer, or as thin as can be conveniently worked, and made rough on both sides with a coarse file, or toothing plane. The veneers of wood or other matter to be combined with them should also be toothed; and, both the plates and veneers being warmed, first pass a coat of glue over one of the metal plates and cover it with a thin sheet of paper, then coat the paper with glue, and cover it with the veneer. Place them between two smooth and even boards, and let them be kept together either by a screw-press, or by hand-screws, and remain till dry; they will then be found to adhere together with sufficient firmness for cutting to the pattern.

The pattern should be drawn on the veneer, or if, from the colour, it should not be sufficiently distinct, a piece of paper may be pasted on the veneer, and after it is dry the design may be drawn upon it. The lines of the pattern should be cut with a bow-saw, having a very thin and narrow blade; such a saw may be made of part of a watch-spring, and the bow, or the stretcher, of the saw, is required to be at such a distance from the blade as will admit the latter to turn and follow the lines of the pattern in any direction. The frame of the saw should be as light as possible. Where the pattern does not in any place approach the edge, a small hole must be made for inserting the saw; and it is usual to saw upwards, that mode of sawing rendering it more easy to follow the lines correctly. When the whole of the pattern is cut out, the veneer or shell may he separated from the metal by exposing them to steam, or to warm water.

The next object is to join the parts so as to produce two complete ornaments; the one composed of veneer inlaid with metal, the other of metal inlaid with veneer. For this purpose, on a plain surface, place a piece of paper of sufficient size, and the veneer upon it, then with strong glue insert the metal-part in the veneer, and rub it well down with the veneering-hammer and glue; next, cover the whole with another piece of paper, and place it between two plain boards, which had been previously well warmed and rubbed with tallow, and screw or press them together. If this be properly done, the work will separate from the boards when dry; and, the paper being removed, it may be laid in its place as a veneer; but a caul is usually employed in preference to the hammer. The reverse pattern, it is obvious, should be prepared for laying in the same manner.

The process is the same whether metal and wood, or metal and tortoiseshell, or two woods of different colours be used.

Inlaying with Shaded Wood.

Having shown the methods of cutting out and veneering, we need now only show the method used to produce that shady brown edge, on works inlaid with white holly, and which, when well executed, has a very pleasing and ornamental effect; the method is as follows:—

Into a shallow iron or tin pot, put a sufficient quantity of fine dry sand, to be level with the top edge of it; place it on the fire till it is quite hot, then having your veneer cut out to the required pattern, dip the edges into the hot sand, and let them remain till the heat has made them quite brown; but be careful not to burn them; it is best to bring them to a proper colour, by repeatedly renewing the operation, than all at once, as you then do not injure the texture of the wood, and by immersing more or less of the edge, you produce a shaded appearance to your satisfaction. I would here recommend the workman, previous to beginning the operation, to have his pattern before him, shaded with umber, or any brown colour, in those parts that the wood is to be stained, as he then will be enabled, as he proceeds, to copy the various shades of the pattern, for the wood when once shaded cannot be altered; and as much of the beauty of this work depends on a proper judgment in placing your shadows, it is best always to have a guide to go by, that we may produce the best possible effect. Sometimes it is requisite to give a shadow in the centre, and not on the edge of your wood; and as this cannot be done by dipping it in the sand, you must do it by taking up a little of the hot sand, and sprinkling it, or heaping it up on those parts required to be darkened, letting it remain a short time, then shaking it off, and, if necessary, apply more where the colour is not deep enough.

To imitate Inlaying of Silver Strings, &c.

This process is sometimes employed in the stocks, &c. of pistols, and if well executed has a very good effect; carefully draw your pattern upon the work, and then engrave, or cut away the different lines with sharp gouges, chisels, &c. so as to appear clean and even, taking care to cut them deep enough, and rather under, like a dovetail, to secure the composition afterwards to be put in the channels. The composition to resemble silver, may be made as follows: take any quantity of the purest and best grain tin, melt it in a ladle or other convenient receptacle: add to it, while in fusion, the purest quicksilver, stirring it to make it incorporate; when you have added enough, it will remain in a stiff paste; if too soft, add more tin, and if not sufficiently fluid, add quicksilver; grind this composition on a marble slab, or in a mortar, with a little size, and fill up the cuttings or grooves in your work, as you would with a piece of putty; let it remain some hours to dry, when you may polish it off with the palm of your hand, and it will appear as if your work was inlaid with silver. Instead of tin, you may make a paste of silver leaf and quicksilver, and proceed as above directed; you may also for the sake of variety in your work, rub in wax of different colours, and having levelled the surface and cleaned off your work, hold it at a moderate distance from the fire, which will give your strings a good gloss.

A superior Glue for the above.

Melt your glue as usual, and to every pint add of finely-powdered rosin and finely-powdered brick-dust two spoonful each; incorporate the whole well together, and it will hold the metal much faster than plain glue.

To Polish Brass Ornaments Inlaid in Wood.

If your brass work be very dull, file it with a small smooth file; then polish it with a rubber of hat dipped in Tripoli powder mixed with linseed oil, in the same manner as you would polish varnish, until it has the desired effect.

To Wash Brass Figures over with Silver.

Take one ounce of aqua-fortis, and dissolve in it over a moderate fire one drachm of good silver cut small, or granulated; this silver being wholly dissolved, take the vessel off the fire, and throw into it as much white tartar as is required to absorb all the liquor. The residue is a paste, with which you may rub over any work made of copper, and which will give it the colour of silver.

To Gild Metal by dissolving Gold.

Dissolve gold in aqua regia, and into the solution dip linen rags; take them out and dry them gently; then burn them to tinder; after you have well polished your work with this, take a cork, and dipping it into common salt and water, and afterwards into the tinder, rub your work well, and its surface will be gilt.

Carving, Reeding, &c.

In carving, the quality of the wood is of the utmost importance. It should be free from cracks, knots, &c., and as even in its texture as possible, and, above all, well seasoned.

The first thing to be done is to draw your pattern on the wood in its proper proportions; this is called boasting, and in it consists the chief art of carving, as he who is the best skilled in drawing, has the best idea of the quantity of projection that should be given to the respective parts, to accord with the given design. After making out the sketch, the carver has to shape the outline with saws or gouges, and then make out the prominences of each part when necessary or proper, by glueing on pieces of wood for that purpose. The roughly-formed pieces are fixed for carving, and, in some cases, this is done by glueing them to a board, with paper inserted between, to enable the carver to take the carving off with more ease when it is finished. When the work is properly fixed, the carver proceeds to place his gouges; and, by a judicious choice of such kinds only as will suit the turn of the parts in boasting, endeavours not to have more than he can use without confusion.

The principal lines of the whole are then formed, so as to be a sufficient guide to finishing, when it is completed with gouges and cutting tools of various kinds.

The union of carved and turned work has almost always a beautiful effect; but, in producing richness with the smallest degree of labour, the combination may be carried to a great extent.

Reeding is a kind of ornament much in use in all parts of turned work. It is far better than fluting or cabling, for it has a bolder effect in small work than in fluting. When reeding is introduced on flat surfaces, there should always be an odd number, as 3, 5, 7, &c., the centre one being a trifle bolder in table legs, bed pillars, &c.

Moulding Ornaments, Figures, &c. in Imitation of Carving.

To avoid the expense of carving in wood, several attempts have been made to cast figures and ornaments to resemble wood. The most approved process we here present our readers. It was invented by M. Lenormand, and rewarded at the Exposition of French products, in 1823.

Make a very clear glue with parts of Flanders glue, and one part of isinglass, by dissolving the two kinds separately in a large quantity of water, and mix them together after they have been strained through a piece of fine linen, to separate the filth and heterogeneous parts which could not be dissolved. The quantity of water cannot be fixed, because all kinds of glue are not homogeneous, so that some require more and some less; but the proper degree of liquidity may be known by suffering the mixed glue to become perfectly cold, it must then barely form a jelly. If it happens that it is still liquid when cold, a little of the water must be evaporated by exposing the vessel in which it is contained to heat. On the other hand, if it has too much consistence, a little warm water must be added. The glue, thus prepared, is to be heated till you can scarcely endure your finger in it; by this operation a little water is evaporated, and the glue acquires more consistence. Then take raspings of wood, or saw-dust, sifted through a fine hair-sieve, and with the glue form it into a paste, which must be put into plaster or sulphur moulds, after they have been well rubbed over with linseed or nut-oil, in the same manner as when plaster is to be moulded. Care must be taken to press the parts into the mould with the hand, in order that the whole may acquire the perfect form: then cover it with an oiled board, place over it a weight, and suffer it in that manner to dry. The drying may be hastened a little, and rendered more complete, by a stove. When the casting is dry remove the rough parts, and if any inequalities remain behind they must be smoothed, and then the ornament may be affixed with glue to the article for which it is intended.

It may be varnished or polished in the usual manner This operation is exceedingly easy; nothing is necessary but moulds, and, with a little art, the ornament may be infinitely varied.

The species of ornament called *Composition Ornament*, is used where the mass is not great, and the surface can be covered with gilding or paint, and is not exposed to wear. Sunk roses, and other ornaments, which are protected by projections or mouldings, may be done in this manner, and it may be successfully applied to all objects beyond the reach of accident.

The composition is made as follows :—Mix 14 pounds of glue, 7 pounds of rosin, $\frac{1}{2}$ pound of pitch, $2\frac{1}{2}$ pints of linseed oil, and 5 pints of water, (more or less according to the quantity required.) Boil the whole together, well stirring till dissolved; adding as much whiting as will render it of a hard consistency; then press it into your mould which has been previously oiled with sweet oil.

No more should be mixed than can be used before it becomes sensibly hard, as it will require steaming before it can be again used.

Composition ornaments should be well glued on, and, in some cases, they will require to be further secured by needle-points or brads.

Composition ornaments are chiefly used for picture and glass frames; we have also seen them employed for ornaments on the top of oak book-cases, and, when grained by a good painter, they answer as well as when carved in wood.

DYING, STAINING, POLISHING, VARNISHING, &c.

Dying wood is mostly applied for the purpose of veneers, while staining is more generally had recourse to, to give the desired colour to the article after it has been manufactured.—In the one case, the colour should penetrate throughout; while in the latter, the surface is all that is essential.

In dying pear-tree, holly, and beech, take the best black; but for most colours holly is preferable.—It is also best to have your wood as young and as newly cut as possible. After your veneers are cut, they should be allowed to lie in a trough of water for four or five days before you put them into the copper; as the water, acting as a purgative to the wood, brings out abundance of slimy matter; which, if not thus removed, the wood will never be of a good colour; after this purifactory process, they should be dried in the open air for at least twelve hours: they are then ready for the copper. By these simple means, the colour will strike much quicker, and be of a brighter hue. It would also add to the improvement of the colours, if, after your veneers have boiled a few hours, they are taken out, dried in the air, and again immersed in the colouring copper. Always dry veneers in the open air; for fire invariably injures the colours.

Fine Black.

Put six pounds of chip logwood into your copper, with as many veneers as it will conveniently hold, without pressing too tight; fill it with water, and let it boil slowly for about three hours; then add half a pound of powdered verdigris, half a pound of copperas, and four ounces of bruised nut-galls; fill the copper up with vinegar as the water evaporates; let it boil gently two hours each day, till the wood is dyed through.

ANOTHER.

Procure some liquor from a tanner's pit, or make a strong decoction of oak-bark, and to every gallon of the liquor add a quarter of a pound of green copperas, and mix them well together: put the liquor into the copper, and make it quite hot, but not to boil; immerse the veneers in it, and let them remain for an hour; take them out, and expose them to the air till it has penetrated its substance; then add some logwood to the solution, place your veneers again in it, and let it simmer for two or three hours; let the whole cool gradually, dry your veneers in the shade, and they will have acquired a very fine black.

Fine Blue.

Into a clean glass bottle, put one pound of oil of vitriol, and four ounces of the best indigo pounded in a mortar; (take care to set the bottle in a basin or earthen glazed pan, as it will ferment; now put your veneers into a copper, or stone trough; fill it rather more than one-third with water, and add as much of the vitriol and indigo (stirring it about) as will make a fine blue, which you may know by trying it with a piece of white paper or wood; let the veneers remain till the dye has struck through.

Fine Yellow.

Reduce four pounds of the root of barberry, by sawing to dust, which put in a copper or brass trough; add four ounces of turmeric, and four gallons of water, then put in as many white holly veneers as the liquor will cover; boil them together for three hours, often turning them; when cool, add two ounces of aqua-fortis, and the dye will strike through much sooner.

Bright Yellow.

To every gallon of water, necessary to cover your veneers, add one pound of French berries; boil the veneers till the colour has penetrated through; add the following liquid to the infusion of the French berries, and let your veneers remain for two or three hours, and the colour will be very bright.

Bright Green.

Proceed as in either of the above receipts to produce a yellow; but instead of adding aqua-fortis or the brightening liquid, add as much vitriolated indigo as will produce the desired colour.

ANOTHER.

Dissolve four ounces of the best verdigris, and half an ounce of indigo, in three pints of the best vinegar; put in your veneers, and gently boil till the colour has penetrated sufficiently.

The hue of the green may be varied by altering the proportion of the ingredients.

Bright Red.

To two pounds of genuine Brazil dust, add four gallons

of water; put in as many veneers as the liquor will cover, boil them for three hours; then add two ounces of alum, and two ounces of aqua-fortis, and keep it lukewarm until it has struck through.

ANOTHER.

To every pound of logwood chips, (well picked from dirt, &c.,) add two gallons of water; put in your veneers, and boil as in the last; then add a sufficient quantity of the brightening liquid (page 55) till you see the colour to your mind; keep the whole as warm as you can bear your finger in it, till the colour has sufficiently penetrated.

The logwood chips is always best when fresh cut, which may be known by its appearing of a bright red colour; for if stale it will look brown, and not yield so much colouring matter.

Purple.

To two pounds of chip logwood and half a pound of Brazil dust, add four gallons of water, and after putting in your veneers, boil them for at least three hours; then add six ounces of pearlash and two ounces of alum; let them boil for two or three hours every day, till the colour has struck through.

The Brazil dust only contributes to make the purple of a more red cast; you may therefore omit it, if you require a deep blush purple.

Orange.

Let the veneers be dyed, by either of the methods given in page 54, of a fine deep yellow, and while they are still wet and saturated with the dye, transfer them to the bright red dye as in page 55, till the colour penetrates equally throughout.

Silver Grey.

Expose to the weather in a cast-iron pot of six or eight gallons, old iron nails, hoops, &c. till covered with rust; add one gallon of vinegar, and two of water, boil all well for an hour; have your veneers ready, which must be air-wood (not too dry,) put them in the copper you use to dye black, and pour the iron liquor over them; add one pound of chip logwood, and two ounces of bruised nut-galls; then boil up another pot of the iron liquor to supply the copper with, keeping the veneers covered, and boiling two hours a day, till of the required colour.

Liquid for Brightening and Setting Colours.

To every pint of strong aqua-fortis, add one ounce of grain tin, and a piece of sal-ammoniac of the size of a walnut; set it by to dissolve, shake the bottle round with the cork out, from time to time; in the course of two or three days it will be fit for use. This will be found an admirable liquid to add to any colour, as it not only brightens it, but renders it less likely to fade from exposure to the air.

STAINING.

STAINING wood is altogether a different process from dying it, and requires no preparation before the stain be applied: it is peculiarly useful to bedstead and chair makers. In preparing the stain, but little trouble is required; and, generally speaking, its application differs very little from that of painting. When carefully done, and properly varnished, staining has a very beautiful appearance, and is much less likely to meet with injury than japanning.

To make Imitation Rosewood.

Brush the wood over with a strong decoction of logwood, while hot; repeat this process three or four times; put a quantity of iron-filings amongst vinegar; then, with a flat, open brush, made with a piece of cane, bruised at the end, or split with a knife, apply the solution of iron-filings and vinegar to the wood in such a manner as to produce the fibres of the wood required. After it is dry, the wood must be polished with turpentine and bees'-wax,

Brown Veined Stain, or Imitation of Rosewood.

For the ground, make a stain by boiling 16 parts of logwood in 64 parts of water, till the liquor acquires a deep red colour. Then add one part of carbonate of potash, and apply the stain hot, letting the work become nearly dry between each coat, till a good rosewood ground be formed; after which let it become quite dry.

To form the veins, heat the black stain above described, as used for the last coat, and with a graining brush, such as is used by painters, make the dark veins on the work. The veins should be disposed to represent the dark parts of the natural wood with as much taste and skill as possible.

Brown Stains to Imitate Mahogany.

The surface may first be rubbed with a diluted solution

of aqua-fortis; then one ounce of dragon's blood being dissolved in a pint of spirit of wine by heat, and one-third of an ounce of carbonate of soda being added, the mixture is filtered, and afterwards laid on with a soft brush. On being done over a second time, the wood acquires the external appearance of mahogany.

To Stain Beech a Mahogany Colour.

Put two ounces of dragon's blood, broken in pieces, into a quart of rectified spirits of wine; let the bottle stand in a warm place, shake it frequently, when dissolved, it is fit for use.

To give any Close-grained Wood the appearance of Mahogany.

The surface of the wood must first be planed smooth, and then rubbed with weak aquafortis; after which it is to be finished with the following varnish :—To three pints of spirits of wine is to be added four ounces and a half of dragon's blood, and an ounce of soda, which have been previously ground together; after standing some time, that the dragon's blood may be dissolved, the varnish is to be strained, and laid on the wood with a soft brush. This process is to be repeated, and then the wood possesses the perfect appearance of mahogany. When the polish diminishes in brilliancy, it may be speedily restored by rubbing the article with a little linseed oil.

To Take Ink out of Mahogany.

Mix in a tea-spoonful of cold water a few drops of oil of vitriol, and touch the spot with a feather dipped in the liquid, taking care neither to exceed nor to come short of the due quantity of acid: too little doing no good, and too much only substituting one stain for another. In the latter case, or where the colour is made lighter than the rest of the wood, perhaps a little linseed oil would be the readiest and best restorative.

Easy Method of Darkening Mahogany.

Nothing more is necessary than to wash the mahogany with lime water, which may readily be made by dropping a nodule of lime into a basin of water.

Imitation of Ebony.

Pale-coloured woods are stained in imitation of ebony by washing them with, or steeping them in, a strong decoction of logwood or galls, allowing them to dry, and then washing them over with a solution of the sulphate or acetate of iron. When dry, they are washed with clean water, and the process repeated, if required. They are, lastly, polished or varnished.

Black Stain, or Imitation of Ebony.

First, form a stain of galls and logwood, in the proportion, by weight, of 12 parts of logwood to 2 parts of galls, and give the work one coat with this stain. Add one part of verdigris to the stain, and give the work another coat. Then add one part of sulphate of iron; and apply one or more coats as may be deemed necessary

Black Stain for immediate use.

Boil half a pound of chip logwood, in two quarts of water, add one ounce of pearl-ash, and apply it hot to the work with a brush. Then take half a pound of logwood, boil it as before in two quarts of water, and add half an ounce of verdigris, and half an ounce of copperas strain it off, put in half a pound of rusty steel filings, with this go over your work a second time.

To Imitate King or Botany-bay Wood.

Boil half a pound of French berries, in two quarts of water, till of a deep yellow, and, while boiling hot, give two or three coats to your work; when nearly dry, form the grain with the black stain, which must also be used not.

You may, for variety, to heighten the colour, after giving it two or three coats of yellow, give one of strong logwood liquor, and then use the black stain as directed.

Red Stain for Bedsteads and Common Chairs.

Archil, as sold at the shops, will produce a very good stain of itself, when used cold; but if, after one or two coats, being applied and suffered to get almost dry, it is brushed over with a hot solution of pearlash in water, it will improve the colour.

To Improve the Colour of any Stain.

Mix in a bottle one ounce of nitric acid, half a tea-spoonful of muriatic acid, a quarter of an ounce of grain tin, and two ounces of rain water. Mix it at least two days before using, and keep your bottle well corked.

To Stain Horn in Imitation of Tortoiseshell.

Mix an equal quantity of quick-lime and red-lead with strong soap lees, lay it on the horn with a small brush, in imitation of the mottle of tortoiseshell; when dry, repeat it two or three times.

How to Weld Tortoiseshell.

Provide yourself with a pair of pincers or tongs, so constructed that you can reach four inches beyond the rivet; then have your tortoiseshell filed clean to a lap-joint, carefully observing that there is no grease about it; wet the joint with water, apply the pincers hot, following them with water, and you will find the shell to be joined as if it were one piece.

To Stain Ivory or Bone Red.

Boil cuttings of scarlet cloth in water, and add by degrees pearlash till the colour is extracted; a little roach alum, now added, will clear the colour; then strain it through a linen cloth. Steep your ivory or bone in aqua-fortis (nitrous acid) diluted with twice its quantity of water; then take it out, and put it into your scarlet dye till the colour is to your mind; be careful not to let your aqua-fortis be too strong, neither let your ivory remain too long in it; try it first with a slip of ivory, and if you observe the acid has just caused a trifling roughness on its surface, take it out immediately, and put it into the red liquid, which must be warm, but not too hot; a little practice, with these cautions, will enable you to succeed according to your wishes; cover the places you wish to remain unstained with white wax, and the stain will not penetrate in those places, but leave the ivory of its natural colour.

To Stain Ivory or Bone Black.

Add to any quantity of nitrate of silver (lunar caustic,)

three times its bulk of water, and steep your ivory or bone in it; take it out again in about an hour, and expose it to the sunshine to dry, and it will be a perfect black.

To Stain Ivory or Bone Green.

Steep your work in a solution of verdigris and sal-ammoniac in weak aquafortis, in the proportion of two parts of the former to one of the latter, being careful to use the precautions mentioned for staining red.

To Stain Ivory, &c. Blue.

Stain your materials green according to the previous process, and then dip them in a strong solution of pearl-ash and water.

To Stain Ivory, &c. Yellow.

Put your ivory in a strong solution of alum water, and keep the whole some time nearly boiling; then take them out and immerse them in a hot mixture of turmeric and water, either with or without the addition of French berries; let them simmer for about half an hour, and your ivory will be of a beautiful yellow. Ivory or bone should dry very gradually, or it will split or crack.

To Soften Ivory.

Slice a quarter of a pound of mandrake, and put in half a pint of the best vinegar, into which put your ivory; let it stand in a warm place for forty-eight hours, you will then be able to bend the ivory to your mind.

To give Wood a Gold, Silver, or Copper Lustre.

Grind about two ounces of white beech sand in a gill of water, in which half an ounce of gum arabic has been dissolved, and brush over the work with it. When this is dry, the work may be rubhed over with a piece of gold, silver, or copper, and it will in a measure assume their respective colours and brilliancy. The work may be polished by a flint burnisher, but should not be varnished.

POLISHING.

THE beauty of Cabinet-work depends upon the care with which it is finished; some clean off with scraping and rubbing with glass paper: this should be done in all cases, but it is not enough, particularly where the grain is any-ways soft: a good glass paper is also essential; (directions for making which will be found in page 20,) a polish should then be added. But unless the varnish for cabinet-work be very clear and bright, it will give a dingy shade to all light-coloured woods; this should therefore be a previous care.

The French Method of Polishing.

With a piece of fine pumice-stone and water, pass regularly over the work with the grain, until the rising of the grain is down; then with powdered Tripoli and boiled linseed oil polish the work to a bright face; this will be a very superior polish, but it requires considerable time.

Cheap Oil Polish.

The cheapest and most simple polish is, first, having well cleared the work, to oil the article with linseed oil, when by oiling and rubbing for a short time a bright gloss will be produced, and the natural colour of the wood will show to much advantage. When it is required to darken the colour, alkanet root, dragon's blood, or other colouring matters which dissolve in oil, slightly heated, are mixed with the above.

To Polish Ivory.

If ivory be polished with putty and water, by means of a rubber made of hat, it will in a short time produce a fine gloss.

To Polish any work of Pearl.

Go over it with pumice stone, finely powdered, (first washed to separate the impurities and dirt,) with which you may polish it very smooth; then apply putty-powder as directed for ivory, and it will produce a fine gloss and a good colour.

To Polish Tortoiseshell or Horn.

Having scraped your work perfectly smooth and level, rub it with very fine sand-paper or Dutch rushes; repeat the rubbing with a bit of felt dipped in very finely powdered charcoal with water, and lastly with rotten-stone or putty powder; and finish with a piece of soft wash leather, damped with a little sweet oil.

French Polishing.

All polishes are used much in the same way. If your work be porous, or the grain coarse, it will be necessary, previous to polishing, to give it a coat of clear size previous to your commencing with the polish; and when dry, gently go over it with very fine glass paper; the size will fill up the pores and prevent the waste of the polish.

by being absorbed into the wood, and be also a saving of considerable time in the operation.

The true French Polish.

To one pint of spirits of wine, add a quarter of an ounce of gum copal, a quarter of an ounce of gum arabic, and one ounce of shellac.

Let your gums be well bruised, and sifted through a piece of muslin. Put the spirits and the gums together in a vessel that can be closely corked; place them near a warm stove, and frequently shake them; in two or three days they will be dissolved: strain it through a piece of muslin, and keep it tight corked for use.

French Polish.

To one pint of spirits of wine add half an ounce of gum-shellac, half an ounce of gum-lac, a quarter of an ounce of gum sandarac; place the whole in a gentle heat, frequently shaking it, till the gums are dissolved, when it is fit for use.

German Polish.

The wood is prepared with pumice-stone rubbed flat, oiled, and then rubbed together till smooth. The only varnish then used is a solution of shel-lac in spirits of wine, the clearest grains of lac being for the lightest varnish. Colour red with Brazil wood, and yellow by turmeric root. It is applied with a rubber of five pieces of linen; the varnish is then put on with sponge, and, having soaked through the linen layers, a little linseed oil is added in the midst of the varnish, and the whole extent of the surface of the article to be polished, must be gone over at once with this rubber.

Before the preceding composition is applied, the furniture should be cleaned with hot beer, and all ink or other stains removed.

An improved Polish.

To a pint of spirits of wine, add, in fine powder, one ounce of seed-lac, two drachms of gum guiacum, two drachms of dragon's blood, and two drachms of gum mastic; expose them in a vessel stopped close, to a moderate heat for three hours, until you find the gums dissolved; strain it into a bottle for use, with a quarter of a gill of the best linseed oil, to be shaken up well with it.

This polish is more particularly intended for dark-coloured woods, for it is apt to give a tinge to light ones, as satin-wood, or air-wood, &c., owing to the admixture of the dragon's blood, which gives it a red appearance.

Waterproof Polish.

Take a pint of spirits of wine, two ounces of gum benzoin, a quarter of an ounce of gum sandarac, and a quarter of an ounce of gum anime; these must be put into a stopped bottle, and placed either in a sand-bath or in hot water till dissolved; then strain it; and after adding about a quarter of a gill of the best clear poppy oil, well shake it up, and put it by for use.

Bright Polish.

A pint of spirits of wine, to two ounces of gum benzoin, and half an ounce of gum sandarac, put in a glass bottle corked, and placed in a sand-bath, or hot water, until you find all the gum dissolved, will make a beautiful clear polish for Tunbridge-ware goods, tea-caddies, &c.: it must be shaken from time to time, and when all dissolved, strained through a fine muslin sieve and bottled for use.

Strong Polish.

To be used in the carved parts of cabinet work with a brush, as in standards, pillars, claws, &c.

Dissolve two ounces of seed-lac and two ounces of white rosin in one pint of spirits of wine.

This varnish or polish must be laid on warm, and if the work can be warmed also, it will be so much the better; at any rate moisture and dampness must be avoided.

Directions for Cleaning and Polishing Old Furniture.

Take a quart of stale beer or vinegar, put a handful of common salt, and a table spoonful of spirits of salt into it, and boil it for a quarter of an hour; you may keep it in a bottle, and warm it when wanted for use; having previously washed your furniture with soft hot water to get the dirt off, wash it carefully with the above mixture; then polish, according to the directions, with any of the foregoing polishes.

Mahogany furniture may be cleaned and improved, by taking three-pennyworth of alkanet root, one pint of cold-drawn linseed oil, and two-pennyworth of rose pink; or a part only of the alkanet and rose pink may be added, if the pinky shade occasioned by them should be disagreeable. These ingredients are put together into a pan, to stand all night: the mixture is then rubbed on tables and chairs, and suffered to remain one hour. After this, it is to be rubbed off with a linen cloth, and it will leave a beautiful gloss on the furniture.

To take Bruises out of Furniture.

Wet the part with warm water; double a piece of brown paper five or six times, soak it, and lay it on the place; apply on that a hot flat-iron till the moisture is evaporated; if the bruise be not gone, repeat the process. After two or three applications, the dent or bruise will be raised level with the surface. If the bruise be small, merely soak it with warm water, and apply a red-hot poker very near the surface; keep it continually wet, and in a few minutes the bruise will disappear.

To make Furniture Paste.

Scrape two ounces of bees' wax into a pot or basin; then add as much spirits of turpentine as will moisten it through; at the same time powder an eighth part of an ounce of rosin, and add to it, when dissolved to the consistence of paste, as much Indian red as will bring it to a deep mahogany colour: stir it up, and it will be fit for use.

Polishing Paste.

Half a pound of mottled soap cut into pieces, mixed with half a pound of rotten-stone in powder; put them into a saucepan, with enough of cold water to cover the mixture (about three pints); boil slowly till dissolved to a paste.

ANOTHER, FOR LIGHT COLOURED WOODS.

Scrape a quarter of a pound of bees' wax into half a pint of turpentine, and mixing with the same about a quarter of a pint of linseed oil. Bees' wax scraped, and set in a warm place, with turpentine enough to make it into a paste, will keep the wood still lighter.

Furniture Oil.

Put linseed oil in a glazed pipkin, with as much alkanet root as it will cover; let it boil gently till it becomes of a strong red colour; let it cool, and it will be fit for use.

ANOTHER.

Take of cold-drawn linseed oil, one pint, into which put one ounce of powdered rose-pink; stir it well together, and add one ounce of alkanet root, beat in a mortar. Let the whole stand in a warm place for a few days, when the oil will be deeply coloured, and, the substances having settled, the oil may be poured off for use.

This is an excellent method of darkening new mahogany.

VARNISHING.

In London it is hardly worth while to make varnish, unless required in large quantities, as there are several shops where it may be had very good, and at a fair price; but in the country, where the carriage is an object, and you cannot depend upon the genuineness of the article, it is necessary to be known by the practical mechanic;[*] yet where it can be purchased, we should recommend it to be had.

When wood, or other porous material, is to be varnished, it ought to be coated with some substance which will

---

[*] There being so many kinds of varnishes, and so variously prepared according to the nature of the work for which it is required, that only those essential to the Cabinet-maker are here given, but we would recommend those desirous of making their own varnishes, to obtain "*The Painters, Grainers, and Writers' Assistant,*" which treats fully on the subject.

glare of eggs, gum-water, or gum tragacanth, are occasionally employed, the object in view being to prevent the absorption of the varnish by a coating of some substance not soluble in spirit. When linseed oil is used, it ought to be rubbed on sparingly, then wiped carefully off, and a day or two should be allowed for it to harden, before the varnish is put on.

Turpentine Varnish.

Take of black resin one pound and a half, oil of turpentine two pints. Melt the resin, and after having removed it from the fire, mix in, gradually, the turpentine. Strain if necessary.

Varnish for Furniture.

Melt one part of virgin's white wax, with eight parts of petroleum; by a slight coat of this mixture on the wood with a fine brush while warm, the oil will then evaporate, and leave a thin coat of wax, which should afterwards be polished with a coarse woollen cloth.

An excellent Varnish for Cabinet Work.

Take shell-lac one ounce and a half, gum mastic and gum sandarac of each half an ounce, spirits of wine twenty ounces. The gums to be first dissolved in the spirit, and lastly the shell-lac; this may be performed by putting the mixture into a bottle loosely corked, and placing it in a vessel of warm water, which must not boil, keeping the bottle in the warm water until the gums are dissolved. Should evaporation take place, an equal quantity of spirits of wine so lost, must be replaced in the bottle; let the whole settle, and pour off the clear liquid for use, leaving the sediment behind, but do not filter it.

To make Gold Varnish.

Take gum-lac, well picked, put it into a small linen bag, and wash it in pure water, till the water becomes no longer red, then take it from the bag and suffer it to dry. When it is perfectly dry, reduce it to a fine powder Then take four parts of spirits of wine, and one of gum, reduced, as before directed, to an impalpable powder, so that for every four pounds of spirits you may have one of gum; mix these together, and having put them into an alembic, graduate the fire so that the gum may dissolve in the spirits. When dissolved, strain the whole through a strong piece of linen cloth; throw away what remains as of no use, and preserve the liquor in a glass bottle, closely corked. This varnish may be employed for gilding any kind of wood.

When you wish to use it, you must, in order that the work may be done with more smoothness, employ a brush made of the tail of a Vari, (to be obtained at all artists' colour shops,) and with this instrument dipped in the liquor, wash over gently, three times, the wood which has been silvered; let each coat dry before the next is applied, and your work will resemble the finest gold.

A Varnish for Wood that will resist Boiling Water.

Take a pint and a half of linseed oil, and boil it in a copper vessel, not tinned, suspending in the oil a small linen bag, (which must not touch the bottom of the vessel) containing five ounces of litharge, and three ounces of minium well powdered, till the oil acquires a deep brown colour; then take out the bag and substitute another containing six cloves of garlic, then throw into the liquid one pound of yellow amber, after being melted in the following manner:—To one pound of well-powdered amber, add two ounces of linseed oil, and place them on a strong fire. When the fusion is complete, pour it boiling hot into the prepared linseed oil, and let it continue to boil for five minutes, stirring it well; let it stand a short time, then pour off the composition and preserve it, when cold, in stoppered bottles; after having polished your wood and given it the required colour; when perfectly dry lay on the varnish with a fine sponge, give your work three or four coats, letting each coat dry before another is applied.

To Varnish a piece of Furniture.

First make the work quite clean; then fill up all knots or blemishes with cement of the same colour; see that your brush is clean, and free from loose hairs; then dip your brush in the varnish, stroke it along the wire raised across the top of your varnish pot, and give the work a thin and regular coat; soon after that another, and another, always taking care not to pass the brush twice in the same place; let it stand to dry in a moderately warm place, that the varnish may not chill.

When you have given your work about six or seven coats, let it get quite hard (which you will prove by pressing your knuckles on it; if it leave a mark, it is not hard enough); then with the three first fingers of your hand rub the varnish till it chafes, and proceed over that part of the work you mean to polish, in order to take out all the streaks, or partial lumps made by the brush; then give it another coat, and let it stand a day or two to harden.

To Polish Varnish.

Put two ounces of powdered Tripoli into an earthen

pot or basin, with water sufficient to cover it; then with a piece of fine flannel four times doubled, laid over a piece of cork rubber, proceed to polish your varnish, always wetting it well with the Tripoli and water; you will know when the process is complete, by wiping a part of the work with a sponge, and observing whether there is a fair and even gloss; clean off with a bit of mutton suet and fine flour.

To keep Brushes in order.

It is necessary to be very careful in cleaning them after being used, for if laid by with the varnish in them, they are soon spoiled; therefore, after using, wash them well in spirits of wine or turpentine, according to the nature of your varnish; after which you may wash them out with hot water and soap, and they will be as good as new, and the spirits that are used for cleaning, may be used to mix with varnish for the more common purposes, or the brushes may be cleaned, merely with boiling water and strong yellow soap.

PART IV.—GILDING.

The art of silvering, as applied to cabinet work, is precisely similar to that of gilding; the directions for the one will therefore be the instructions for the other, with little other variation than using silver leaf instead of gold leaf.

There are two methods of gilding:—that for out-door work, to stand the weather, or to wash, is called oil gilding; this is performed by means of oil or varnish.

The other, called burnish-gilding, is the most beautiful, and best adapted for fine work, as frames, articles of furniture, &c. or as applied by the cabinet-maker, in the internal decoration of rooms, or the carved work of its furniture.

The materials to be provided with.

First, a sufficient quantity of leaf gold, which is of two sorts, the deep gold, as it is called, and the pale gold; the former is the best; the latter very useful, and may occasionally be introduced for variety or effect.

Second, a gilder's cushion; an oblong piece of wood, covered with rough calf-skin, stuffed with flannel several times doubled, with a border of parchment, about four inches deep at one end, to prevent the air blowing the leaves about when placed on the cushion.

Thirdly, a gilding knife, with a straight and very smooth edge, to cut the gold.

Fourthly, several camel-hair pencils in sizes, and tips, made of a few long camel's hairs put between two cards, in the same manner as hairs are put into tin cases for brushes, thus making a flat brush with a very few hairs. Lastly, a burnisher, which is a crooked piece of agate set in a long wooden handle.

Size for Oil Gilding.

Grind calcined red ochre with the best and oldest drying oil, and mix with it a little oil of turpentine when used.

When you intend to gild your work, first give it a coat of parchment size; then apply the above size where requisite, either in patterns or letters, and let it remain till by touching it with your fingers it feels just sticky;

then apply your gold leaf, and dab it on with a piece or cotton; in about an hour wash off the superfluous gold with sponge and water; and, when dry, varnish it with copal varnish.

A Size for preparing Frames.

To half a pound of parchment shavings, or cuttings of white leather, add three quarts of water, and boil it in a proper vessel till reduced to nearly half the quantity; then take it off the fire, and strain it through a sieve: be careful in the boiling to keep it well stirred, and do not let it burn.

To prepare Frames or Wood-work.

First, with the above alone, and boiling-hot, go over your frames in every part; then mix a sufficient quantity of whiting with size, to the consistency of thick cream, with which go over every part of your frame six or seven times, carefully letting each coat dry before you proceed with the next, and you will have a white ground fit for gilding on, nearly or quite the sixteenth of an inch in thickness.

Your size must not be too thick, and when mixed with the whiting should not be put on so hot as the first coat is by tself: it will be better to separate the dirty or coarse parts of the whiting, by straining it through a sieve.

Polishing.

When the prepared frames are quite dry, clean and polish them; to do this, wet a small piece at a time, and with a smooth fine piece of cloth dipped in water, rub the part till all the bumps and inequalities are removed, and for those parts where the fingers will not enter, as the mouldings, &c. wind the wet cloth round a piece of wood, and by this means make the surface all smooth and even alike.

Where there is carved work, &c. it will sometimes be necessary to bring the mouldings to their original sharpness, by means of chisels, gouges, &c. as the preparation will be apt to fill up all the finer parts of the work, which must be thus restored; it is sometimes the practice, after polishing, to go over the work, once, with fine yellow or Roman ochre, but this is rarely necessary.

Gold Size.

Grind fine bol-ammoniac well with a muller and stone: scrape into it a little beef suet, and grind all well together; after which, mix in with a pallet knife a small proportion of parchment size with a double proportion of water.

ANOTHER.

Grind a lump of tobacco-pipe clay into a very stiff paste with thin size; add a small quantity of ruddle, and fine black lead ground very fine, and temper the whole with a small piece of tallow.

To Prepare your Frames for Gilding.

Take a small cup, or pipkin, into which put as much gold size as you judge sufficient for the work in hand, add parchment size, till it will just flow from the brush; when quite hot, pass over your work with a very soft brush, taking care not to put the first coat too thick; let it dry, and repeat it twice or three times more, and when quite dry, brush the whole with a stiff brush, to remove any remaining nobs. Your work is now ready for applying the gold.

Your parchment size should be of such a consistence, when cold, as the common jelly sold in the shops; for if too thick it will be apt to chip, and if too thin it will not have sufficient body.

Laying on the Gold.

This is the most difficult part of the operation, and requires some practice; but with a little caution and attention, it may be easily performed.

Turn your gold out of the book on your cushion a leaf at a time; then passing your gilding knife under it, bring it into a convenient part of your cushion for cutting it into the size of the pieces required; breathe gently on the centre of the leaf, and it will lay flat on your cushion, then cut it to your mind by bringing the knife perpendicularly over it, and sawing it gently till divided.

Place your work before you in a position nearly horizontal, and with a long-haired camel-hair pencil, dipped in water (or with a small quantity of brandy in the water,) go over as much of your work as you intend the piece of gold to cover; then take up your gold from your cushion with your tip; by drawing it over your forehead, or cheek, it will damp it sufficiently to adhere to the gold, which must then be carefully transferred to your work, and gently breathing on it, it will adhere; but take care that the part you apply it to is sufficiently wet; indeed, it must be floating, or you will find the gold apt to crack; proceed in this manner by a little at a time, and do not attempt to cover too much at once, till by experience you are able to handle the gold with freedom. Be careful, in proceeding with your work, if you find any flaws, or cracks appear, to take a corresponding piece of gold, and apply it immediately.

Burnishing.

When your work is covered with gold, set it by to dry, it will be ready to burnish in about eight or ten hours; but it will depend on the warmth of the room or state of the air, and practice will enable you to judge of the proper time.

When it is ready, those parts which you intend to burnish must be dusted with a soft brush, and wiping your burnisher with a piece of soft wash-leather (quite dry), begin to burnish about an inch or two in length at a time, taking care not to lean too hard, but with a gentle and quick motion apply the tool till you find it equally bright all over.

Matting or Dead Gold.

Those parts of your work which look dull from not being burnished, are now to be matted, that is, are to be made to look like dead gold; for if left in its natural state it will have a shining appearance, which must be thus rectified:—

Grind some vermillion, or yellow ochre, very fine, and mix a very small portion either with the parchment-size or with the white of an egg, and with a very soft brush lay it even and smooth on the parts intended to look dull; if well done, it will add greatly to the beauty of the work.

Finishing.

It is now only necessary to touch the parts in the hollows with a composition made by grinding vermillion, gamboge, and red lead, very fine, with oil of turpentine, and applying it carefully with a small brush in the parts required, and your work is completed.

Sometimes the finishing is done by means of shell gold, which is the best method; it should be diluted with gum arabic, and applied with a small brush.

To make Shell Gold.

Take any quantity of leaf gold, and grind it, with a small portion of honey, to a fine powder; add a little gum arabic and sugar-candy, with a little water, and mix it well together; put it in a shell to dry against you want it.

An excellent Receipt to Burnish Gold Size.

One ounce of blacklead, ground very fine, one ounce of deer suet, one ounce of red chalk, and one pound of pipe-clay, ground with weak parchment-size to a stiff consistency, to be used as directed in the article 'Size for oil gilding.'

To Clean Oil Paintings.

Clean the picture well with a sponge, dipped in warm beer; after it has become perfectly dry, wash it with a solution of the finest gum-dragon, dissolved in pure water. Never use blue starch, which tarnishes and eats out the colouring; nor white of eggs, which casts a thick varnish over pictures, and only mends bad ones by concealing the faults of the colouring.

Gold Varnish for Leather.

Take of turmeric root and gamboge, of each one scruple and a half; of oil of turpentine, two pints; seed-lac and gum juniper, four ounces of each; dragon's blood, half an ounce; Venice turpentine, two ounces; and clean sand, four ounces. Well mix and pour off the clear solution.

To Gild Leather for Bordering Doors, Folding Screens, &c.

Damp a clear brown sheep-skin with a sponge and water, and strain it tight, with tacks, on a board sufficiently large; when dry, size it with clear double-size; then beat the whites of eggs, with a whisk, to a foam, and let them stand to settle; then take books of leaf silver, a sufficient quantity, and blow out the leaves of silver on a gilder's cushion; pass over the leather carefully with the egg size, and with a tip brush lay on the silver, closing any blister with a bit of wool; when dry, varnish them over with yellow lacker till they are of a fine gold colour. Your skin being thus gilt, you may then cut it into strips as you please, and join with paste to any length.

To Gild the Borders of Leather Tops of Library Tables, Work Boxes, &c.

The tops of library tables, &c. are usually covered with Morocco leather, and ornamented with a gilt border, and are usually sent to the bookbinder for that purpose. The method by which they perform it is as follows:—They first go over that part intended to be gilt with a sponge dipped in the glare of eggs, which is the whites beaten up to a froth, and left to settle; and the longer made or older it is, so much the better; then being provided with a brass roller, on the edge of which the pattern is engraved and fixed as a wheel in a handle, they place it before the fire till heated, so that, by applying a wetted finger, it will just hiss; while it is heating, rub the part with an oiled rag, or clean tallow, where the pattern is intended to be, and lay strips of gold on it, pressing it down with cotton: then with a steady hand run the roller along the edge of the leather, and wipe the superfluous gold off with an oiled rag, and the gold will adhere in those parts where the impression of the roller has been, and the rest will rub off with the oiled rag.

To make Paste for laying the Cloth or Leather on Table-tops, Desks, &c.

Take one pound of the best wheaten flour, one large spoonful of powdered alum, and two spoonfuls of finely-powdered rosin, well mix them in a pan, then add cold water by degrees, carefully stirring it till it is of the consistence of thin cream, put it into a saucepan over a clear fire, keeping it constantly stirred from the bottom that it may not burn or get lumpy; as soon as it simmers it is done enough; empty it out into a pan, and cover it over till cold to prevent a skin forming at the top, which will make it lumpy.

To Paint Sail Cloth,

So as to render it Pliant, Durable, and Waterproof.

Grind ninety-six pounds of English ochre with boiled oil, and add to it sixteen pounds of black paint. Dissolve one pound of yellow soap in one pail of water on the fire, and mix it, while hot, with the paint. Lay this composition, without wetting it, upon the canvas, as stiff as can conveniently be done with the brush, so as to form a smooth surface; the next day, or the day after, lay on a second coat of ochre and black, with a very little, if any, soap; allow this coat a day to dry, and then finish the canvas with black paint.

Oil Gilding,

Where the object is to give a high finish.

Paint the work with a color composed of the finest white lead and yellow ochre, in such proportions that the color shall be as near as possible to the color of the gold to be employed, mixed with oil (not boiled,) and turpentine, till of the consistence of thin paint; this should be laid on evenly, and allowed to dry thoroughly, then repeat it till it is perceived that the grain or roughness of the object to be gilt is entirely hidden. When the last coat is dry it must be rubbed perfectly smooth, first with a piece of pumice stone, and finished with a piece of woollen cloth and finely pounded pumice; and lastly, with putty powder, till it is smooth as glass. It must then be varnished over with fine lac varnish several times, applying a slight degree of heat after each coat to make the varnish flow smoothly over the surface. When the last coat of varnish is quite hard it must be polished; this is done by putting on a horse-hair glove, and rubbing the surface with this first, then with tripoli, applied with a piece of wet woollen cloth; and lastly, by wet putty powder, first applied with woollen cloth, then with the bare hand, till it is as bright as glass. It must then be varnished over with a thin coat (the thinner the better) of gold size and when sufficiently dry the gold is to be applied, beginning at the part that is dryest. When gilt, it is to be allowed to remain for two or three days, and then brushed over lightly with a camel's hair brush to remove superfluous gold. It is next to be varnished with *spirit varnish*, applying heat as before, then varnished with copal varnish two or three times, allowing it to become perfectly hard between each coat; after the last coat of varnish it is finished by polishing, first with tripoli, applied with a soft cloth and water, and then with the bare hand and a little oil, and wiped dry.

To Gild Oil Painted Work.

If the paint is quite dry and hard, merely paint on the design in *gold size*, and then apply the gold leaf, which must be done carefully. If the paint is not dry around the part to be gilt, it must be dusted over with whitening, and rubbed over with a cloth or brush till the finger will not adhere, before the gold size is applied, and then proceed as above. If the gilding is for out-door work, it must not be varnished, as the sun acting on the varnish gives it a jagged appearance; but if for inside work, it may be varnished over with spirit varnish, and heated slightly by holding a hot iron near it till the varnish has flowed smooth and even over the surface.

To Imitate Mahogany.

Let the first coat of painting be white lead, the second orange, and the last burnt umber or sienna; imitating the veins according to your taste and practice.

To Imitate Wainscot.

Let the first coat be white, the second half white and half yellow ochre, and the third yellow ochre only. Shadow with umber or sienna.

To Imitate Satin-Wood.

Take white for your first coating, light blue for the second, and dark blue or dark green for the third.

To Stain Musical Instruments, &c.

A FINE CRIMSON STAIN.

Boil one pound of ground Brazil, in three quarts of water for an hour; strain it, and add half an ounce of cochineal; boil it again gently for half an hour, and it will be fit for use. If you would have it more of the scarlet tint, boil half an ounce of saffron, in one quart of water, and pass over the work previous to the red stain. Observe, the work must be very clean, and of air-wood, or good sycamore without blemish; when varnished it will look very rich.

FOR A PURPLE STAIN.

One pound of chip logwood, boiled in three quarts of water, for an hour; then add four ounces of pearl-ash, and two ounces of indigo (pounded), and you will have a good purple.

FOR A FINE BLACK.

In general, when black is required in musical instruments, it is produced by japanning, the work being well prepared with size and lampblack; take some black japan and give it two coats, after which varnish and polish it.

A FINE BLUE STAIN.

Put one pound of oil of vitriol in a clean glass bottle, into which put four ounces of pounded indigo, (take care to set the bottle in a basin or glazed earthen pan, as it will ferment); after it is quite dissolved, provide an earthen or wooden vessel, so constructed that it will conveniently hold the article you wish to dye; fill it rather more than one-third with water, into which pour as much of the vitriol and indigo till you find the whole to be a fine blue dye; put in the article and let it remain till the dye has struck through.

A FINE GREEN STAIN.

Take three pints of the strongest vinegar, to which put four ounces of the best verdigris, ground fine; half an ounce of sap-green; and half an ounce of indigo.

FOR A BRIGHT YELLOW.

There is no need to stain the wood, as a very small piece of aloes put in the varnish will make it a good color, and have the desired effect.

Curious mode of Silvering Ivory.

Immerse a small slip of ivory in a weak solution of nitrate of silver, and let it remain till the solution has given it a deep yellow color; then take it out and immerse it in a tumbler of clear water, and expose it in the water to the rays of the sun. In about three hours the ivory acquires a black color; but the black surface, on being rubbed, soon becomes changed to a brilliant silver.

Instructions for Repairing Paintings.

DAMAGED SURFACE.

When by the continued pressure of some hard body, the canvas presents either a concavity or convexity in a portion of its surface, it must be well wet in that part, and left gradually to dry in some cool place, keeping it constantly under pressure.

TO MAKE THE COLORS ADHERE WHEN BLISTERED, &c.

When the color has separated from the priming, whilst the priming still remains firm, the swollen and detached part is first rubbed over with the same paste which will be presently mentioned as used for lining. Then, with a pin or needle, little holes are punctured in the part, and more paste rubbed over these holes with a pencil, and worked about so that it shall pass through them. The surface is then wiped clean, and over the spot a pencil is passed that has been dipped into linseed-oil: this serves to soften it. A warm iron is then passed rapidly over the raised surface, which attaches itself to the priming as before. Should it be necessary to line the canvas with a new one, it should be done previously.

REPAIRING OIL PICTURES.

When a canvas is broken, rent, or perforated in any part, the piece of canvas that is used to repair the damage is dipped into melted wax, and applied the moment it is taken out, warm as it is, to the part, which has been previously brought together as well as possible, and also saturated with the wax. With great care you flatten down the piece; so that as the wax chills and concretes, the parts adhere and are kept smooth. The whole being made perfectly level, and the excess of the wax removed, a mastic made of white lead mixed with starch is applied; for oil-color does not adhere well to wax. The white is afterwards colored thin, or by washes, according to the tone of the surrounding parts, and re-painted.

LINING AND TRANSFERING.

When the priming of a canvas has become detached, or the cloth is so old as to need sustaining, it is customary to line the picture.—But if the canvas is greatly injured, the painting itself is transferred to a new subjectile. In order to render the old canvas and the color softer and more manageable, expose the picture for several days to damp. When all is ready, the first step is to fasten, by a thin flour-paste, white paper over the whole painted side of the picture, to prevent the colors scaling off. Having a new canvas duly stretched on a strong frame; a uniform coat of well-boiled paste, made of rye-flour with a clove of garlic, is spread nicely over it by means of a large brush. With despatch, yet care, a coat of the same paste is spread likewise on the back of the picture.—The latter is then laid upon the new cloth, the two pasted sides, of course, together. With a ball of linen the usual rubbing is given with a strong hand, begining at the centre, and passing to the edges, which must be carefully kept in place the while. In this way, the air is expelled, which remaining would cause blisters.

The picture thus lined is then placed upon a smooth table, the painted side down, and the back of the new canvas is rubbed over boldly with any suitable smoothing-instrument, such as is used for linen, paper, or the like; and a warm iron is then passed over the picture, having on the other side a board to resist the pressure. The paste being heated by this iron, penetrates on the side of the picture, and fixes still more firmly the painting, while on the other side the redundant part of the paste escapes through the tissue of the new cloth, so that there remains everywhere an equal thickness. The iron must not be too hot, and before applying it, several sheets of paper shoud be interposed between it and the paper that was at first pasted on the painting, and which would not be sufficient.

When the lined picture is sufficiently dry, the paper last mentioned is damped, by passing over it a sponge moistened with tepid water. It soon detaches, and with it is removed the paste that secured it to the picture. All that remains is to clean the painting, and where needed to restore it.

The above operation will not, of course, be attempted by the amateur, except for experiment upon some picture of little worth; for even practised hands frequently injure what they were employed to preserve.

To Clean Tapestry, &c.

Let the article be well beat, and freed from dust; fasten it down to a smooth board; mix half a pint of bullock's gall with two gallons of soft water, scrub it well with soap and the gall mixture; let it remain till dry, and it will be perfectly cleansed, and the colors restored to their original brightness; the brush you use should not be hard, and of rather long hairs, or you will damage your work.

FINIS.

INDEX.

	PAGE.
Badger Softeners	10, 39, 40
,, Poonah	40
Bears' Hair Varnish Brushes	36
Blenders' Sable	23
Blowing Apparatus	21
Bordering Brushes	40
Burnishers	32
Camel-hair Pencils	27, 28, 29
Candlestick	24
Caning Tubes	33
Colors, Oil	30
Color Mills	22
Combs—Steel, Leather, &c.	11, 12
,, for dividing Over-grainers	13
Copper-bound Ground Brushes	16
,, Ground Distemper	17
Cotton Wool	27
Dabbers	31
Dippers	25
Dotters—Maple, Camel-hair	11
Dusters	16
Flat French Tools	19, 37
Flat and Round Sables, in Tin	38
Floggers	13
Gilders' Tips	31
,, Cushions	32
,, Knives	32
Gold Size	26
,, Leaf	27
Graining Rollers	15
Ground Brushes	16
Hog-hair Softeners	11
Horn Combs	11
India Rubber Combs	12
Japanners' Pencils	28
Knives, Palette	19
,, Stopping	19
,, Trowel	20
,, Chisel	20
,, Putty	20
,, Hacking	20
Leather Combs	12
Lining Tools	14, 19
Maple Eye Shaders	11
Mahl Sticks	25
Marking Brushes	40
Mops, in Quill	31

	PAGE
Mottlers, Hog-hair, Extra Double Thick	3
,, ,, Double Thick	4
,, ,, Single Thick	4
,, ,, Chisel Edge	4
,, Camel-hair	5
,, ,, Burnt Edges	5
,, ,, Feather ,,	5
,, Hog-hair Wave	5
,, ,, Wavy	6
Oak Over-combing Tools	14
Over-grainers, Hog-hair, Thin	6
,, ,, Thick	7
,, ,, in Wood	7
,, ,, in Knots	8
,, ,, in Tubes	9
,, Sable, in Tubes	9
,, ,, Solid	10
,, Camel-hair, in Tubes	9
,, Fitch	12
Palettes	25
Paper Hangers' Rollers	21
,, Scissors	20
Parchment Cuttings	27
Pencil Cases, Round and Oval	26
Pipe Clay	27
Poonah Brushes	40
Quilled Tools	34
Round Badger Softeners	40
Round and Flat Hog-hair Tools	37
Sable Writers and Liners	23, 24
Sables, in Tin	38
Sash Tools	17
Scene Painters' Tools	34, 35
,, Colors	35
Steel Combs	11
Stenciling Brushes	17
Stippling ,,	18
Tar Brushes	17
Thumb Pieces	12
Tissue Paper	27
Varnish Brushes	35, 36
,, ,, Flat Camel-hair	39
,, ,, Round	39
Veining Horns	12
,, Fitches	14
Wall Liners, Hog-hair	13
Writers' Boxes	29

GRAINERS' BRUSHES, &c.

No. 1.—EXTRA DOUBLE THICK HOG-HAIR MOTTLERS.

The Extra Thick, Double Thick, and Single Thick Mottles, are all used for Maple.

The best method to preserve these Tools is to wash them in clean water, after working, and let them dry thoroughly, as the hairs have a tendency to decay, and, as when used, the hairs are found to break off and spoil the work.

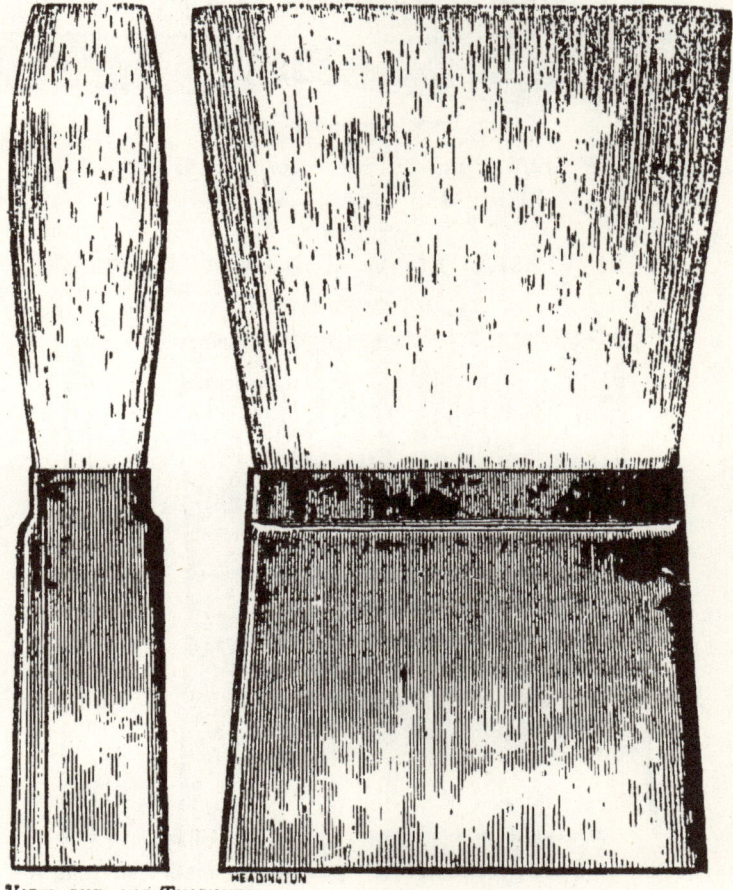

VIEW, SHEWING THICKNESS.

7/-	10/6	14/-	17/6	21/-	24/6	28/-	35/-	42/-	dozen.
1	1½	2	2½	3	3½	4	5	6	inch.

No. 2.—DOUBLE THICK HOG-HAIR MOTTLERS.

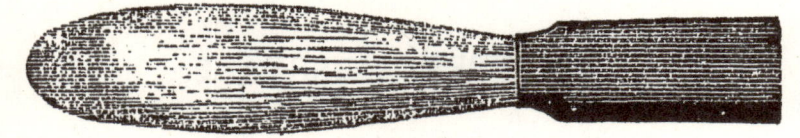

END VIEW, SHEWING THICKNESS.

6/-	9/-	12/-	15/-	18/-	21/-	24/-	27/-	30/-	36/-	dozen.
1	1½	2	2½	3	3½	4	4½	5	6	inch.

No. 3.—SINGLE THICK HOG-HAIR MOTTLERS.

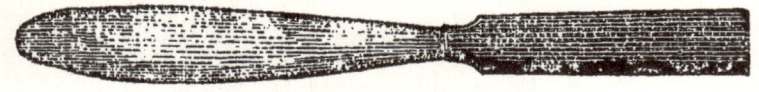

END VIEW, SHEWING THICKNESS.

5/-	7/6	10/-	12/6	15/-	17/6	20/-	22/6	25/-	30/-	dozen.
1	1½	2	2½	3	3½	4	4½	5	6	inch.

No. 4.—CHISEL EDGE HOG-HAIR MOTTLERS,
OR CUTTERS.

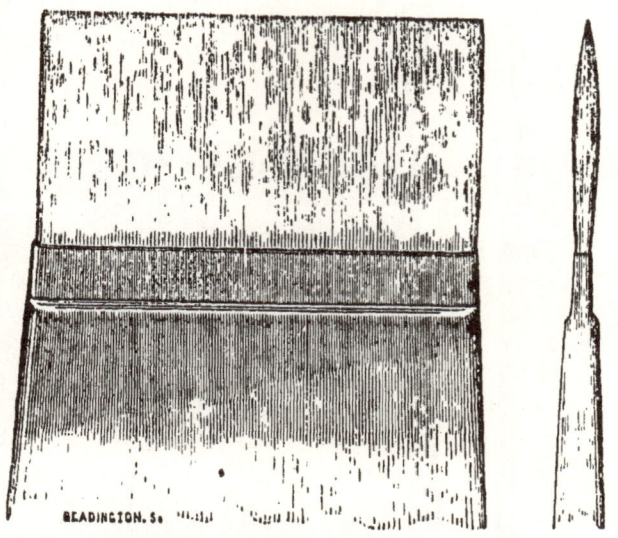

Chiefly used for Satinwood and Mahogany—the smaller sizes as eye tools.

3/-	4/6	5/-	7/6	10/-	12/6	15/-	17/6	20/-	dozen.
½	¾	1	1½	2	2½	3	3½	4	inch.

No. 5.—CAMEL-HAIR MOTTLERS.

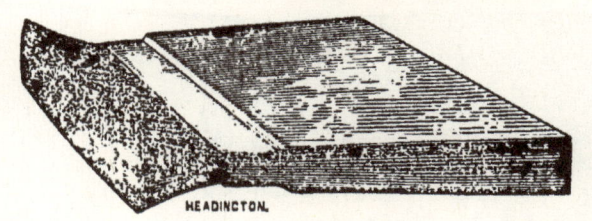

| 6/- | 9/- | 12/- | 15/- | 18/- | 21/- | 24/- dozen. |
| 1 | 1½ | 2 | 2½ | 3 | 3½ | 4 inch. |

Used chiefly for soft woods.

No. 6.—CAMEL-HAIR MOTTLERS,
BURNT EDGES.

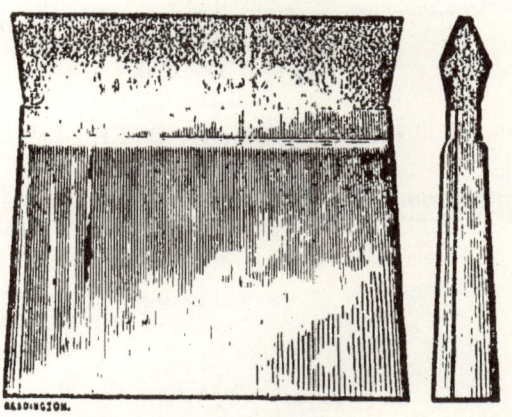

| | 9/- | 12/- | 15/- | 18/- | 21/- | 24/- dozen. |
| 1 | 1½ | 2 | 2½ | 3 | 3½ | 4 inch. |

No. 7.—CAMEL-HAIR MOTTLERS,
FEATHER EDGE.

| 6/- | 9/- | 12/- | 15/- | 18/- | 21/- | 24/- dozen. |
| 1 | 1½ | 2 | 2½ | 3 | 3½ | 4 inch. |

No. 8.—WAVE MOTTLERS (HOG-HAIR).

| 10/6 | 14/- | 17/6 | 21/- | 24/6 | 28/- dozen. |
| 1½ | 2 | 2½ | 3 | 3½ | 4 inch. |

No. 9.—WAVY GRAINERS (HOG-HAIR).

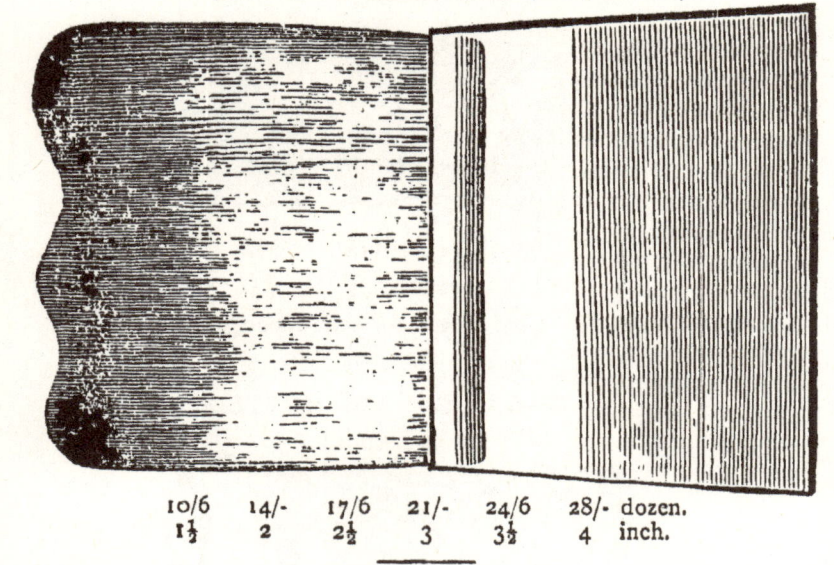

10/6	14/-	17/6	21/-	24/6	28/- dozen.
1½	2	2½	3	3½	4 inch.

No. 10.—THIN HOG-HAIR OVER-GRAINERS, IN TIN.

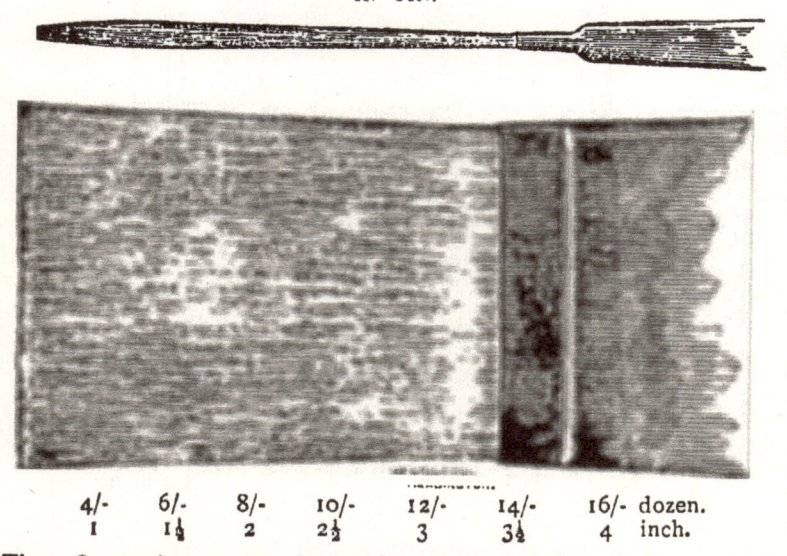

End View.

4/-	6/-	8/-	10/-	12/-	14/-	16/- dozen.
1	1½	2	2½	3	3½	4 inch.

These Overgrainers are much used for Maple, and can also be used for overgraining any wood where fine lines are required. With a comb, the hair may be divided into very fine pencil points. To preserve, see Fig. **1**.

No. 11.—THICK HOG-HAIR, OR OAK, OVER-GRAINERS.

IN TIN.

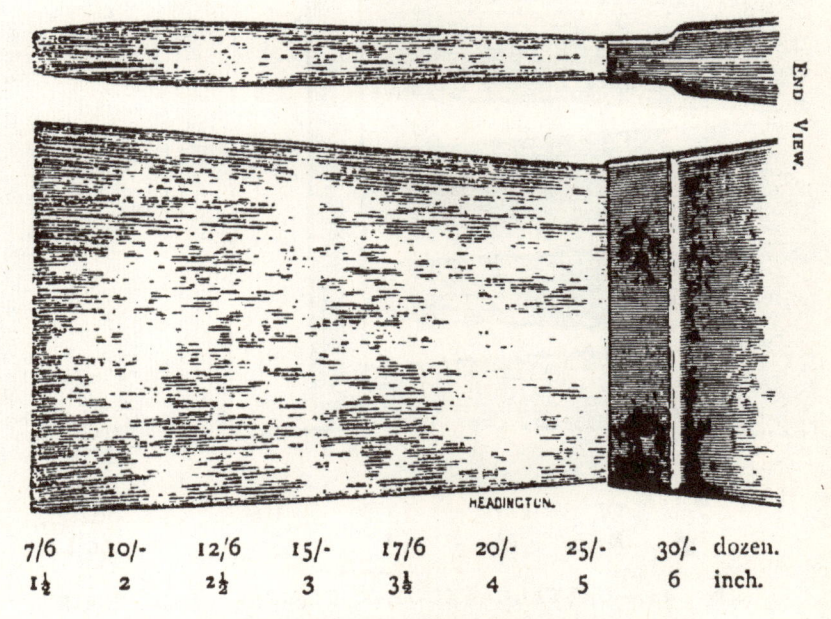

END VIEW.

HEADINGTON.

7/6	10/-	12/6	15/-	17/6	20/-	25/-	30/-	dozen.
1½	2	2½	3	3½	4	5	6	inch.

To preserve the Over-grainers—see note under fig 1.

No. 12.—OAK OVER-GRAINERS,

IN WOOD.

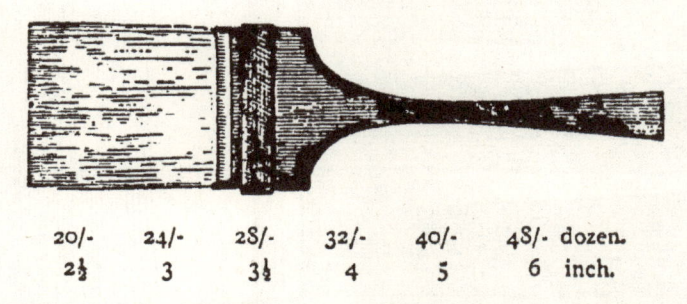

20/-	24/-	28/-	32/-	40/-	48/-	dozen.
2½	3	3½	4	5	6	inch.

To preserve, see note under fig 1, at the same time preventing the handle getting too wet.

No. 13.—OAK OVER-GRAINERS,
IN KNOTS.

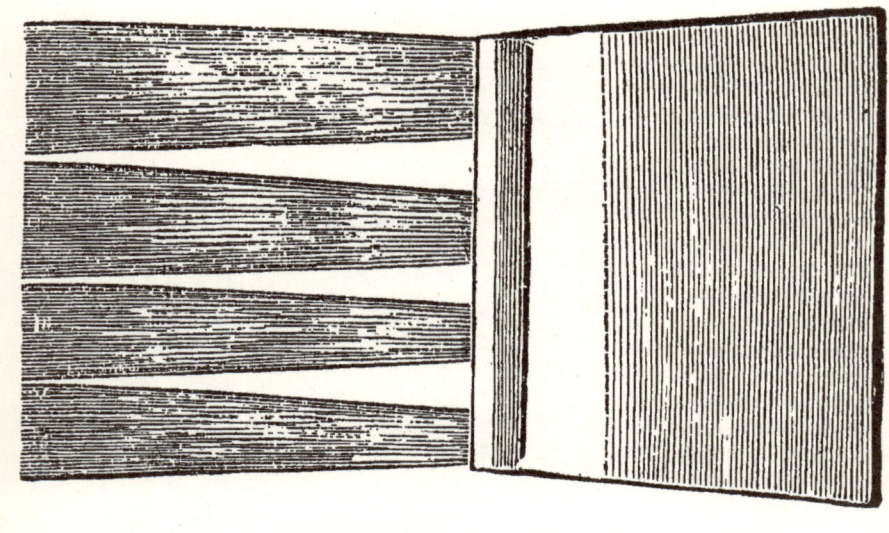

7/	9/	11/6	13/6	16/	18/ dozen.
1½	2	2½	3	3½	4 inch.

No. 14.—CASTELLATED OAK STRIPERS (HOG-HAIR.)

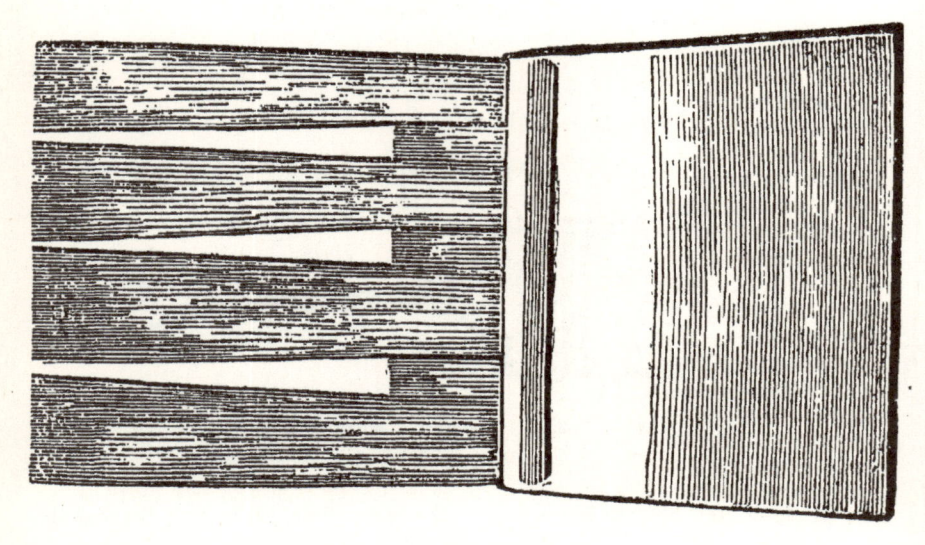

10/	12/6	15/	17/6	20/ dozen.
2	2½	3	3½	4 inch.

To preserve, see note under fig. 1.

No. 15.—HOG-HAIR OVER-GRAINERS,

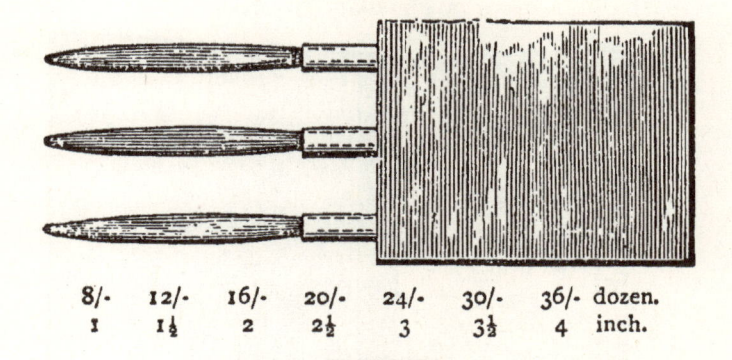

IN TUBES.

8/-	12/-	16/-	20/-	24/-	30/-	36/-	dozen.
1	1½	2	2½	3	3½	4	inch.

No. 16.—SABLE OVER-GRAINERS,

IN TUBES.

Used generally for Maple.

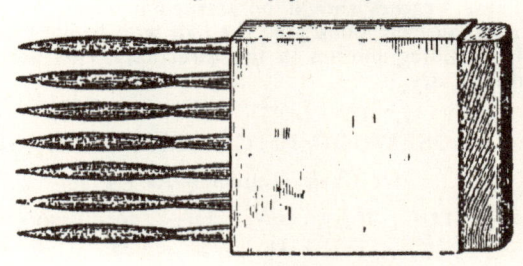

	1	1½	2	2½	3	3½	4	inch
Four Tubes to the inch—	18/-	27/-	36/-	45/-	54/-	63/-	72/-	doz.
Five ,, ,, —	21/-	31/6	42/-	52/6	63/-	73/6	84/-	,,
Six ,, ,, —	24/-	36/-	48/-	60/-	72/-	84/-	96/-	,,

To keep them in working order, they should be washed and then wiped on a nearly dry sponge, and kept dry with the pencils pointed.

No. 17.—CAMEL-HAIR OVER-GRAINERS,

In Tubes—as Fig 16.

	1	1½	2	2½	3	3½	4	inch
Five Tubes to the inch—	12/-	18/-	24/-	30/-	36/-	42/-	48/-	doz.
Four ,, ,, —	9/-	13/6	18/-	22/6	27/-	31/6	36/-	,,

No. 18.—SABLE OVER-GRAINERS,
THICK OR SOLID.
Used generally for Maple.

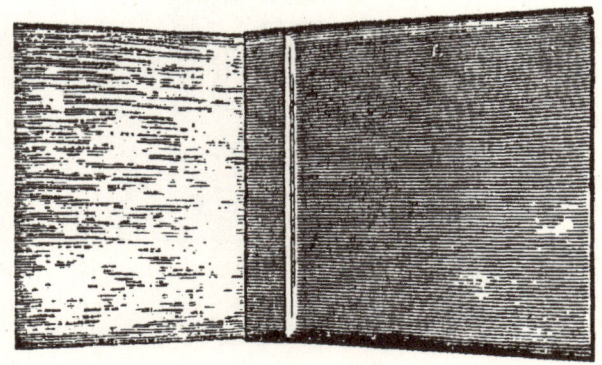

With a comb, the hair may be divided into very fine pencil points.

18/-	27/-	36/-	45/-	54/-	63/-	72/-	dozen.
1	1½	2	2½	3	3½	4	inch.

NOTE.—These Brushes should be washed in clean water—then combed, and laid on a flat surface, smoothing the hair with back of the comb; then lift the brush with a sliding motion in the direction of the hair, in order to keep its shape and let it dry.

No. 19.—ROSEWOOD FITCH HAIR OVER-GRAINERS,
Or Chair Grainers—As Fig 18.

8/-	12/-	16/-	20/-	24/-	30/-	36/-	dozen.
1	1½	2	2½	3	3½	4	inch.

No. 20.—BADGER-HAIR SOFTENERS, OR BLENDERS,
BEST QUALITY.

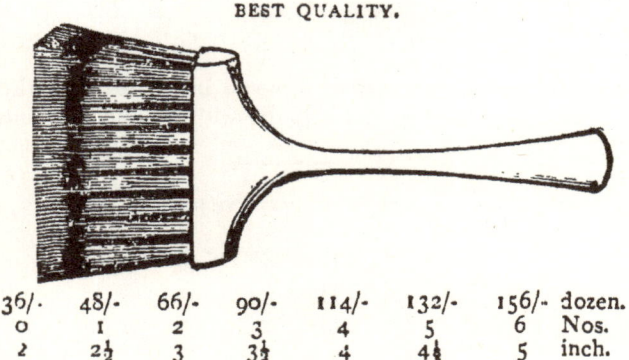

36/-	48/-	66/-	90/-	114/-	132/-	156/-	dozen.
0	1	2	3	4	5	6	Nos.
2	2½	3	3½	4	4½	5	inch.

No. 21.—HOG-HAIR SOFTENERS, OR BLENDERS,
FOR MARBLE.

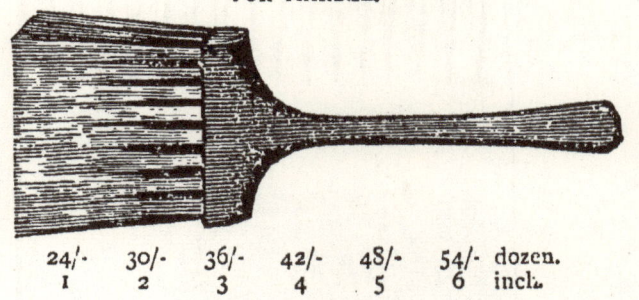

24/-	30/-	36/-	42/-	48/-	54/- dozen.
1	2	3	4	5	6 inch.

No. 22.—MAPLE EYE DOTTERS.

Small, 3/- doz.; Middle, 3/6 doz.; Large, 4/- doz.

No. 23.—MAPLE EYE SHADERS (HOG-HAIR).

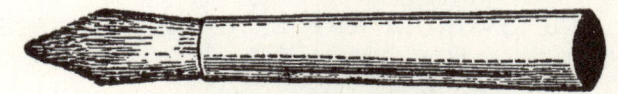

3/- doz., either Small, Middle, or Large.

No. 24.—STEEL GRAINING COMBS.

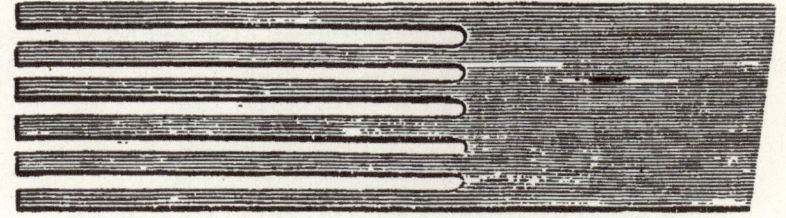

1, 1½, 2, 2¼, 3, 3½, 4, 4½, 5, and 6 inch.   1½d. per inch.
Per Set of 3 each 1, 2, 3, 4, inch, in Case, 3/6.
The above Combs are cut 6, 9, 12, and 15 Teeth to the inch.

No. 25.—HORN COMBS.

1/6	2/6	3/-	4/-	4/6	5/6	6/-	7/6	9/- dozen.
1	1½	2	2½	3	3½	4	5	6 inch.

No. 26.—LEATHER COMBS.

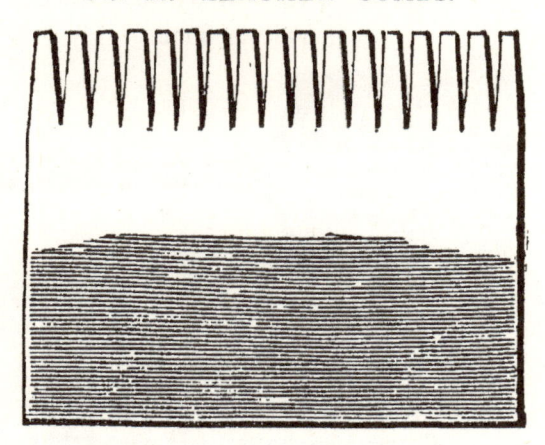

Graduated	.	.	.		4/6	5/-	6/-	dozen.
Coarse, Middle, or Fine,	2/6	3/-		3/6	4/-	4/6	,,
	2	2½		3	3½	4	inch.

These are made of stout Leather, cut by Machinery.

No. 27.—INDIA RUBBER GRAINING COMB,
IN TIN SOCKETS,

Is the only Comb that will Comb Oak in Water Color, and is most useful in connection with the PATENT GRAINING ROLLERS.

By this Comb a much more natural and correct imitation is produced than anything hitherto attained, being superior to ordinary Oil Combing.

They are made in sizes 2, 3, 4, 5, and 6 inches. The teeth cut as follows :—

No. 1 fine, about 12 to the inch.
,, 2 medium ,, 9 ,, ,,
,, 3 coarse 5 ,, ,,   } 4d. per inch.
,, 4 graduated . . . .
,, 5 Irregular . . .

No. 28.—THUMB PIECES, OR VEINING HORNS.

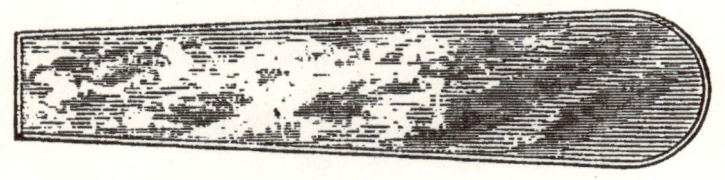

Small, 1/6 doz.; Middle, 4-inch, 2/- doz.; Large, 2/6 doz.

BRODIE & MIDDLETON, 79, LONG ACRE, LONDON.

No. 29.—WALL LINERS (HOG-HAIR).

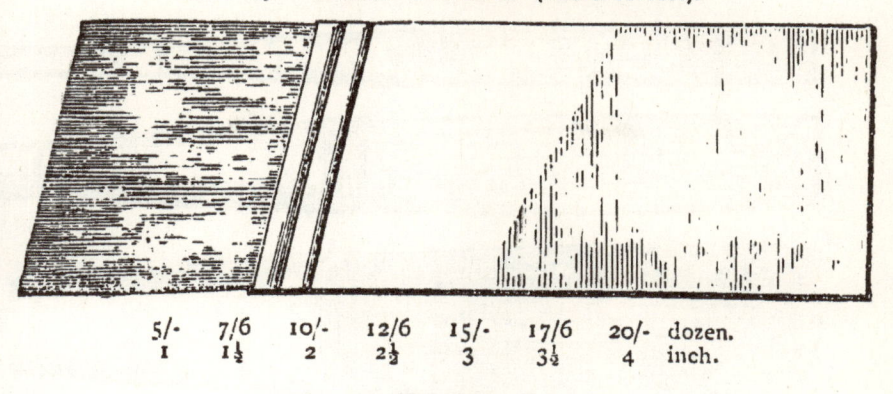

| 5/- | 7/6 | 10/- | 12/6 | 15/- | 17/6 | 20/- | dozen. |
| 1 | 1½ | 2 | 2½ | 3 | 3½ | 4 | inch. |

GRAINERS' COLORS, prepared ready for use, 2/6 lb.

No. 30.—FLOGGING, OR, OAK MATTING TOOLS.

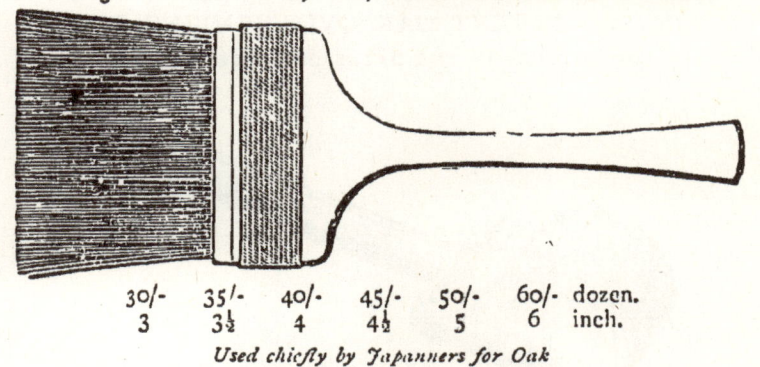

| 30/- | 35/- | 40/- | 45/- | 50/- | 60/- | dozen. |
| 3 | 3½ | 4 | 4½ | 5 | 6 | inch. |

Used chiefly by Japanners for Oak

No. 31.—COMBS FOR DIVIDING OVER-GRAINERS.

Single, 5/- doz. Double, 6/- doz.

The single combs are 6 inches long, and have the same teeth throughout as in the engraving. The double are 7 inches long, one-half being fine and the other coarse, like a dressing comb, but the teeth are cut in such a manner as to cause the hair to divide with a sharp edge at the points.

No. 32.—LINING OR VEINING FITCHES.

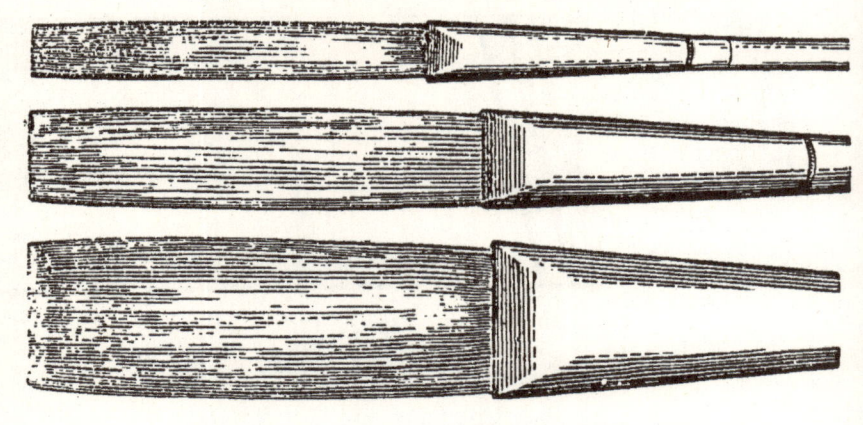

The above three cuts represent No. 3, 6, 9.

No.	1	2	3	4	5	6	7	8	9	10	12
	2/6	2/6	3/-	3/-	3/6	4/-	4/6	5/-	6/-	7/-	8/- doz.

No. 33.—PATENT OAK OVER-COMBING TOOLS.
FOR PUTTING IN THE BATES OR PORES OF THE WOOD.

SEPARATE ROLLERS . . ½-in., 1/9; 2-in., 3/3; 3½-in., 4/- each.

With a 3-in. Feeding Brush fitted into a box. Price 20/- the set of six pieces.

INSTRUCTIONS.—The graining is mixed thinner than for the combing process. Make it so as to flow easily over the work. When the work is rubbed in even, soft, and clean, take a rag and wipe off across the joints of the styles sharply,—to show such an amount of contrast as is seen in the joints of real oak doors. When this is done, proceed to light or figure the door as if it had been combed in the old way, taking care to keep the edges soft by brushing gently either across or along them. The next day, or when the door is dry, it is ready to roll. Mix the colour for rolling thicker than that for graining, using your own judgment as to depth of colour. When this is done, dip your flat brush or feeder into the rolling colour, holding the roller in one hand and the feeder in the other. The feeder is then placed firmly over—and a little behind—the roller, which being set in motion feeds itself. The roller is to be run lightly over the work, but the brush to be kept moderately firm on the roller, resting the hand with the feeder on the other to keep them steady, always commencing at the bottom of the article to be rolled. When the rollers are not in use, keep them either in oil or graining colour.

No. 34.—PATENT GRAINING ROLLERS.

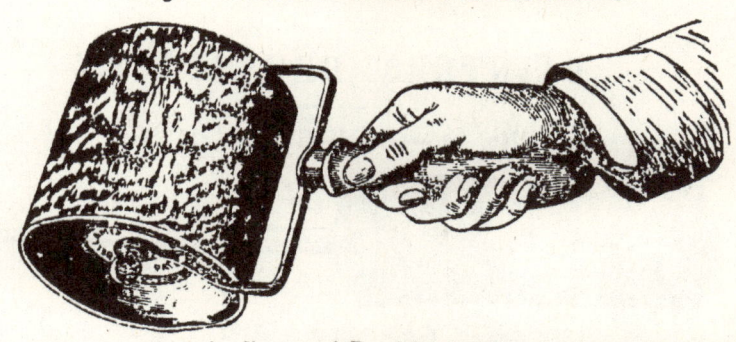

The apparatus consists of a Frame and Revolving Cylinder, the figure of the wood to be represented being cut on the surface of the Cylinder. The designs are of the choicest description. The Apparatus is very durable and expeditious, and has been found to give universal satisfaction. The construction of the Tools are made so as to Grain work of any length or width. Full printed particulars and directions for use will be sent with the Tools.

A description of the WOODS and MARBLES for which the PATENT GRAINING TOOLS are applicable:—

SPANISH	MAHOGANY and Satin-wood Mottles	BIRCH Mottles
Ditto	Heart over-grain	BIRCH Heart Over-grain
Ditto	Feathers	TULIP
MAPLE	Mottles	HUNGARIAN ASH
OAK	Light Veins	HAIR WOOD
Ditto	Dark Veins	JASPER MARBLE
Ditto	Hearts	ROUGE ROYAL MARBLE
Ditto	Knotted and Pollard	VEIN MARBLE, for White, Dove, Sienna, Black and Gold, and Italian Pink
WALNUT		

For further particulars see the Painters' and Grainers' Hand Book, 2s.

Price, 3-in., 16/-; 4-in., 18/-; 5-in., 22/-; 6-in., 26/- each. Mahogany Feathers, 26/-, 30/-, and 37/- each.

LARGER SIZES MADE TO ORDER.

Any of the above Tools will answer for Panels and Style Work generally: smaller sizes are also kept in Stock for Styles, Margins, Frames, Small Panels, Furniture, and all small portions of Woodwork generally.

Price 10/- and 12/- each.

INSTRUCTIONS FOR USING THE PATENT GRAINING TOOLS,
IN WATER COLOR ONLY.

The principal thing to be observed is Cleanliness. Wet them well with sponge and water; have a wash leather wetted, and wring it nearly dry, then spread it on any flat surface ready to run the tool over, to clean it, as it gets charged with colour in using; then lay on your colour mixed in beer in the usual way, on the painted wood work, and pass over it with the tool which produces the figure of the wood it is to represent; then soften it with a Badger-hair Brush. This being done, the work is ready for Over-graining.

THE FOLLOWING WILL BE FOUND A READY AND CLEAN SYSTEM OF GRAINING WORK GENERALLY.

First Grain all your Panels, Styles, Frames, &c., leaving all Mouldings, Edges, &c. to be done last, using for that purpose a small thin one inch and a half Hog-hair Mottler.

Be careful to clean the tools, and not allow them to remain packed inside each other while wet, as they will last much longer if kept dry when not in use.

PAINTERS' BRUSHES, &c.

No. 35.—GROUND BRUSHES.

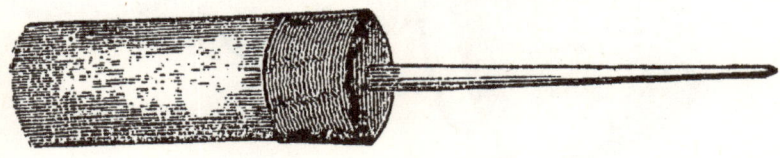

	8-0	6-0	4-0	3-0	2-0	1-0	
Lily Hair—	54/-	48/-	42/-	39/-	34/-	32/-	dozen.
Grey Middles—	48/-	42/-	36/-	34/-	30/-	28/-	,,

Copper 3/- extra.

No. 36.—OVAL GROUND BRUSHES.
STRING BOUND.

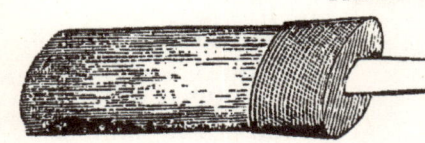

	8-0	6-0	4-0	3-0	
Lily Hair—	59/-	52/-	44/-	42/-	dozen.
Grey Middles—	51/-	45/-	39/-	37/-	,,

Copper 3/- extra.

No. 37.—COPPER-BOUND OVAL GROUND BRUSHES.

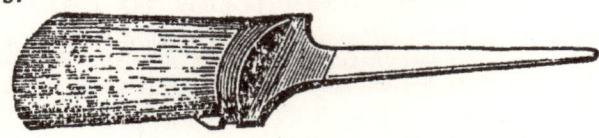

	8-0	6-0	4-0	3-0	
Lily Hair—	60/-	55/-	47/-	45/-	dozen.
Grey Middles—	54/-	48/-	42/-	40/-	,,

No. 38.—DUSTERS.

	8-0	6-0	4-0	3-0	2-0	
6 and 6½-inch Hair—	52/-	45/-	34/-	32/-	28/-	dozen.
7-inch ,,	54/-	48/-	36/-			,,

No. 39.—COPPER-BOUND GROUND DISTEMPER BRUSHES.

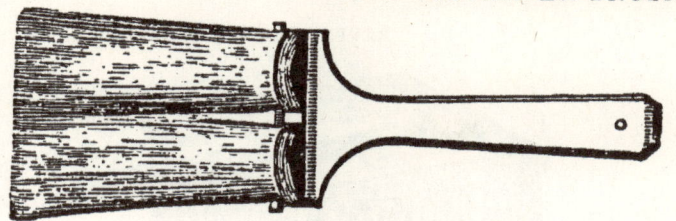

	8-oz.	9-oz.	10-oz.
6-inch Hair—	60/-	66/-	72/- dozen.
6½ ,,	63/-	69/-	75/- ,,

Unground, 3/- per dozen less.

No. 40.—SASH TOOLS.
STRING BOUND.

No.	0	1	2	3	4	5	6	7	8	9	10	12
	2/-	2/6	3/-	4/-	5/-	6/-	8/-	10/6	13/-	16/-	20/-	24/- doz.

No. 41.—TAR BRUSHES.

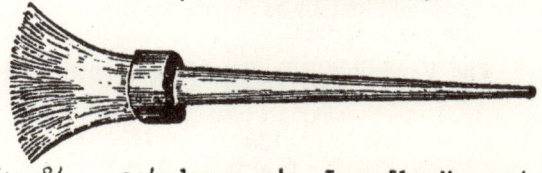

Short Handle.—8/-	14/- dozen.	Long Handle.—12/-	16/- dozen.
No. 1	2	No. 1	2

No. 42.—BEST STENCILLING BRUSHES,
FOR DECORATORS.

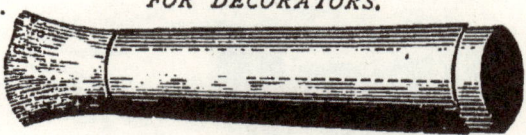

No.	4	6	8	10	12	14	16	18	20	22	24	26
	2/-	3/-	4/-	5/-	6/-	7/-	8/-	10/-	12/-	15/-	18/-	21/- doz.

No. 43.—ROUND STENCILLING BRUSHES.
SET IN KNOTS.
FOR DECORATORS.

2/-	3/6	5/-	7/-	9/- each.
2	2½	3	3½	4 inch.

No. 44.—STIPPLING BRUSHES,

REVERSIBLE HANDLE.

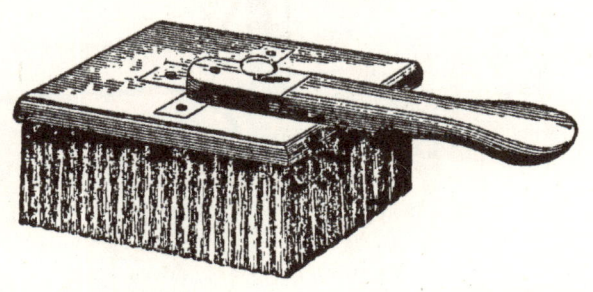

The above cut represents a Stippler, the plan of which, when cnce known, will recommend itself to every painter.

Reversible Stipplers, on this principle, have now been in use for some years past by a celebrated Decorator, who claims to be the originator of the design.

The above handle will be found superior to all others, it being firmer than any of the Reversible Stippler Handles now being used.

The advantages are :—

1.—By loosening the thumb screw, the handle can be turned in any position, and by so doing, the Brush can be evenly worn over its whole surface.

2.—The Reversible Stippler will last much longer than those with the fixed handle.

10/6	13/-	18/-	each.
5 × 5	6 × 6	7 × 7	inches.

With Plain Screw, 1/- each less.

5/-	7/6	10/-	13/-	16/6	each.
5 + 3	6 × 4	7 × 5	8 × 6	9 × 7	inches.

No. 45.—LONG-HAIRED FLAT FRENCH TOOLS.
Same sizes as Fig. 1.

No.	1	2	3	4	5	6	7	8	9	10	12
	3/-	3/-	4/-	4/-	4/6	5/-	6/-	7/-	8/-	9/-	10/- doz.

No. 46.—HOG-HAIR LINING TOOLS,
FOR DECORATORS.

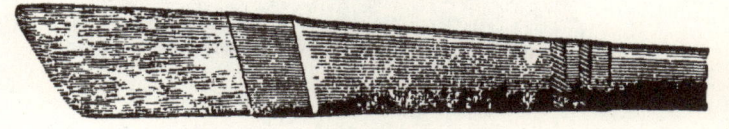

Same sizes as Fig 1.

No.	1	2	3	4	5	6	7	8	9	10	12
	3/-	3/-	4/-	4/-	4/6	5/-	6/-	7/-	8/-	9/-	10/- doz.

No. 47.—STONE KNIVES.

4	5	6	7	8	9	10	11	12	13	14	15	16 inch.
6d.	7d.	8d.	11d.	1/2	1/6	2/-	2/6	3/-	3/6	4/-	4/9	5/3 each.

No. 48.—PAINTERS' PUTTY OR STOPPING KNIVES,
EBONY HANDLE.

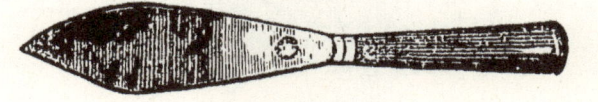

4	4½	5	5½	6 inch.
9/-	10/-	12/-	14/-	16/- doz.

No. 48a.—PLAIN COCOA HANDLE.

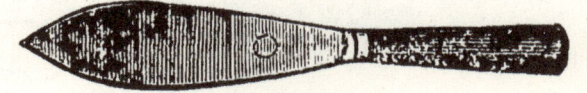

4	4½	5	5½	6 inch.
8/-	9/-	10/-	12/-	14/- dozen.

No. 49.—TROWEL STOPPING KNIVES.

12/-	14/-	16/-	18/-	dozen.
4½	5	5½	6	inch.

No. 50.—CHISEL KNIVES.

10/-	12/-	14/-	16/-	dozen.
4½	5	5½	6	inch.

No. 51.—GLAZIERS' PUTTY KNIVES.

Best Plain, 9/- dozen.   Notched, 10/6 dozen.

No. 52.—GLAZIERS' HACKING KNIVES.

Small, 7/- dozen; Large, 10/- dozen.

No. 53.—PAPER HANGERS' SCISSORS.

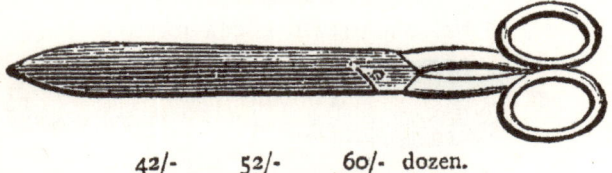

42/-	52/-	60/-	dozen.
10	11	12	inch.

No. 54.—PAPER HANGERS' ROLLER.

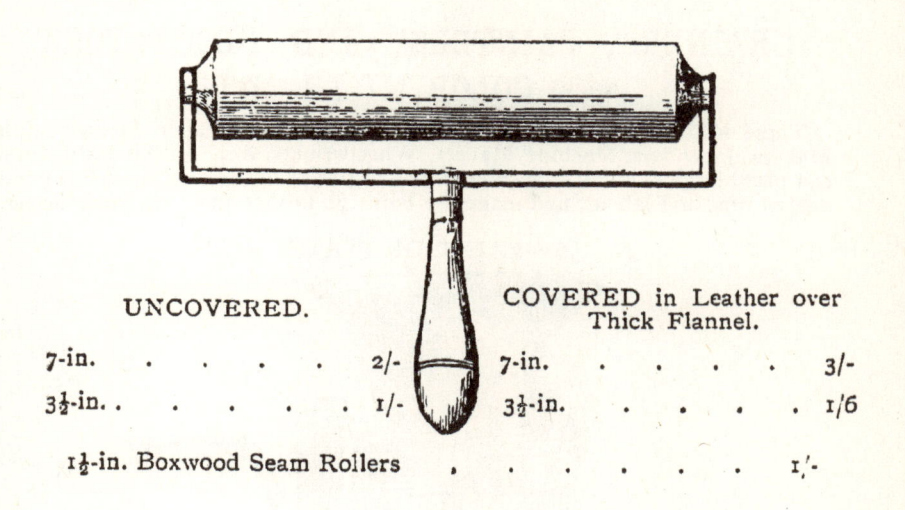

UNCOVERED.		COVERED in Leather over Thick Flannel.	
7-in.	2/-	7-in.	3/-
3½-in.	1/-	3½-in.	1/6

1½-in. Boxwood Seam Rollers 1/-

No. 55.—PATENT SELF-ACTING BLOWING APPARATUS,

For Gas Fitters, Plumbers, Braziers, Painters, &c.

A Safety Valve is adapted to each Apparatus

DIRECTIONS FOR THE MANAGEMENT OF THE SELF-ACTING BLOWING APPARATUS.

Half fill the Boiler with Methylated Spirit, and fill the Lamp with the same, using a piece of common lamp cotton for the Wick, in about half a minute or a minute from the time of lighting the blast will be emitted, and may be directed to any required point. Keep the Wick well in front of the jet to prevent the blast from jumping. By pulling up the Wick a stronger blast may be obtained.

To extinguish it, blow gently on the flame of the lamp from beneath.

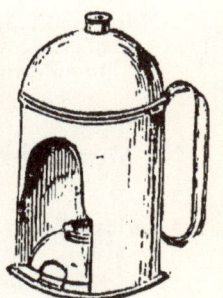

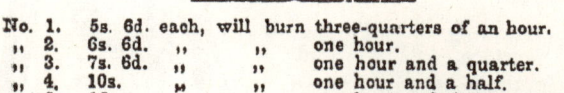

No. 1.	5s. 6d.	each,	will burn	three-quarters of an hour.
,, 2.	6s. 6d.	,,	,,	one hour.
,, 3.	7s. 6d.	,,	,,	one hour and a quarter.
,, 4.	10s.	,,	,,	one hour and a half.
,, 5.	15s.	,,	,,	one hour and three-quarters.

Extra Boilers for Burning off Paint, 3s. to 7s. 6d. each, extra.

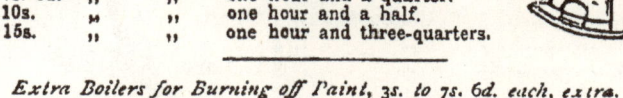

GRAINERS', PAINTERS', AND DECORATORS' COLOR MILLS.

These Mills are used by Painters, Coach Builders, Paper Stainers, Ink Makers, Japanners, Machine Makers, Wheelwrights, &c. In these Mills you can grind large or small quantities to any degree of fineness—they save a great deal of time and labour, and make the Paint go further and look much better.

No. 56.—PAINT OR COLOR MILL.

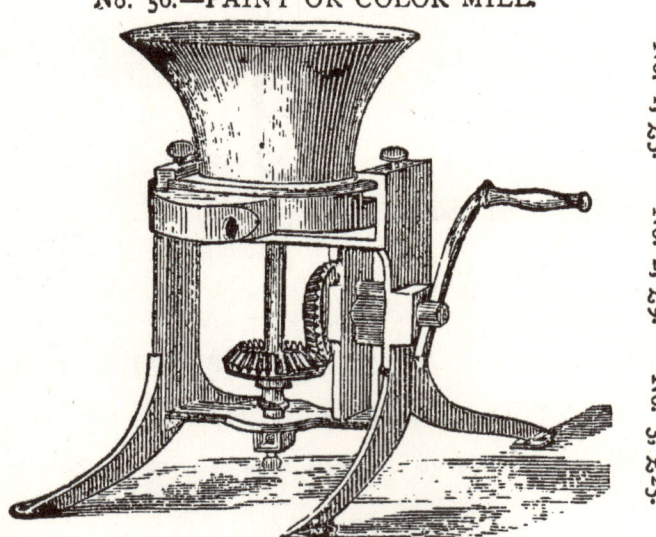

No. 1, £5. No. 2, £9. No. 3, £25.

Improved Paint Mill.	Grinding surface.	Hopper to hold.
No. 1	7-in. diameter	1 Gallon.
„ 2	10 „	3 „
„ 3, for Power	18 „	10 „

No. 57.—AMERICAN PAINT OR COLOR MILL.

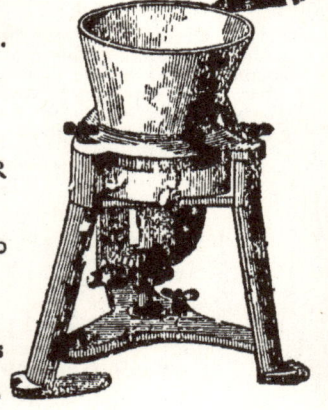

American Paint Mill.	Grinding surface.	Hopper to hold.
No. 1a	5-in. diameter	1 Quart.
„ 2	7 „	4 „
„ 3	10 „	3 Gallons

No. 1a, £4.   No. 2, £5 10s.   No. 3, £9 10s.

WRITERS' AND COACH PAINTERS' BRUSHES, &c.

No. 58.—SABLE-HAIR WRITING PENCILS.

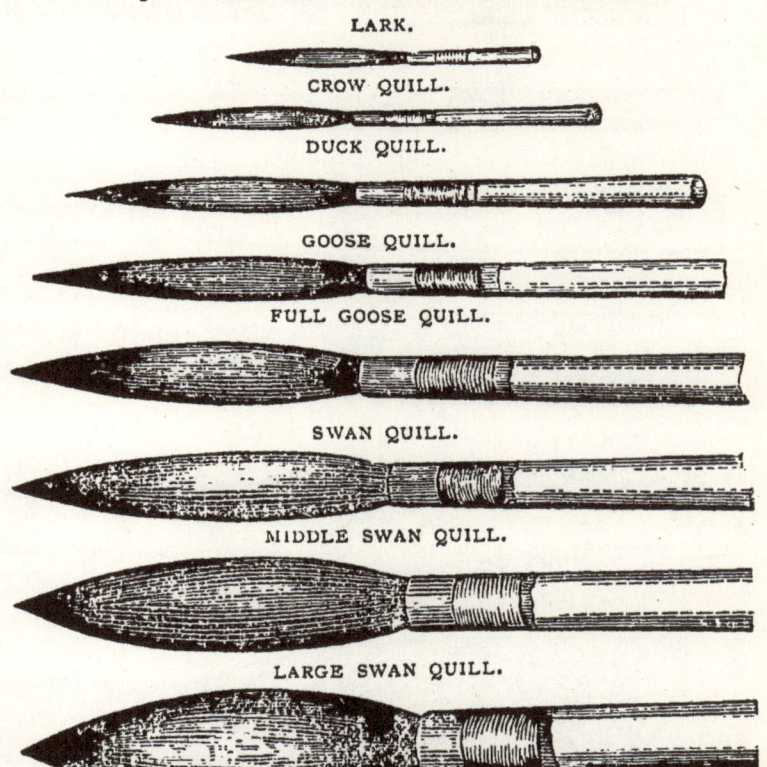

LARK.

CROW QUILL.

DUCK QUILL.

GOOSE QUILL.

FULL GOOSE QUILL.

SWAN QUILL.

MIDDLE SWAN QUILL.

LARGE SWAN QUILL.

Lark, 3/-; Crow, 4/-; Duck, 6/-; Goose, 9/-; Extra Goose, 15/-; Small Swan Quill, 27/-; Middle Swan Quill, 36/-; Large Swan Quill, 48/- per doz.

No. 59.—WRITERS' BLENDERS (SABLE).

1/-, 1/4, 1/8, 2/6, 3/-, 3/6, and 4/- each.
Nos. 1.   2.   3.   4.   5.   6.   7.

No. 60.—SABLE LINERS, OR TRACING PENCILS.

FINE.

CROW QUILL.

DUCK QUILL.

GOOSE QUILL.

SWAN QUILL.

MIDDLE SWAN QUILL.

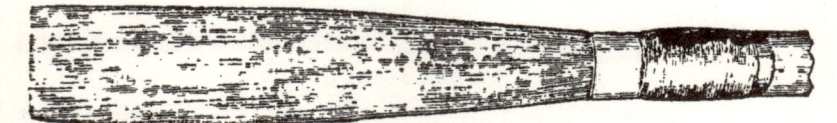

EXTRA LARGE SWAN QUILL.

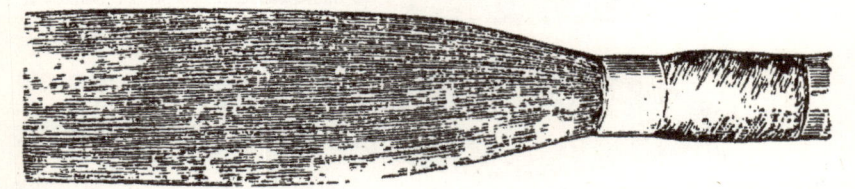

Fine, 3/-; Crow, 4/-; Duck, 6/-; Goose, 9/-; Swan, 27/-; Middle Swan, 36/-; Extra Large Swan, 60/- per doz.
Intermediate sizes can be had at 12/-, 15/-, 18/-, 24/-, 30/- per doz.

No. 61—CANDLESTICK.

To fix on palette or stand on table, with reflector to slide, for working at night.

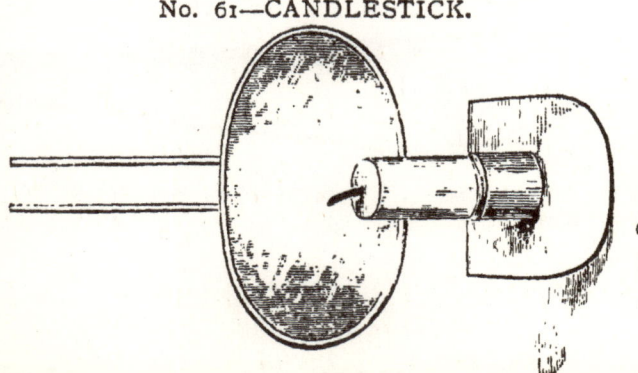

FOR WRITERS, DECORA-TORS, &c.
One Shilling each.

No. 62.—MAHL, OR REST STICK,
From 6d. each.   Jointed from 10d. each.

No. 63.—FOLDING MAHL STICKS,
3/- each.

No. 64.—MAHOGANY PALETTE BOARDS,
OBLONG AND OVAL.

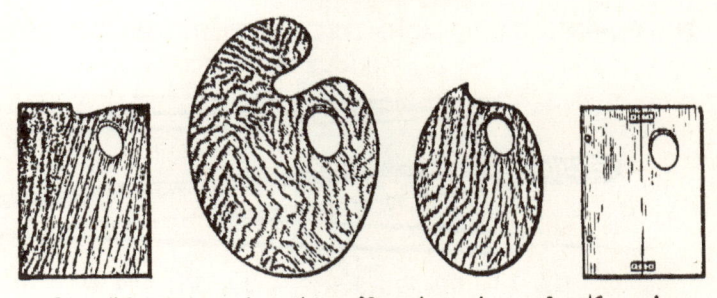

6d., 8d., 9d., 1/-, 1/4, 1/6, 1/9, 2/-, and 2/6 each.

No. 65.—DIPPERS.

		PLAIN.		JAPAN'D				PLAIN.		JAPAN'D	
		s.	d.	s.	d.			s.	d.	s.	d.
No. 1	Single Tin Dippers	0	2½	0	6	No. 9	Double Dippers with Capped Lid	1	6		
,, 2	Double ditto	0	5	1	0	,, 10	Single Conical ditto	0	8	1	0
,, 3	Single Conical ditto	0	2½	0	6	,, 11	Double ditto ditto	1	4	2	8
,, 4	Double ditto ditto	0	5	1	0	,, 12	Single Shallow Tin Dippers	0	3	0	6
,, 5	Single ditto with Neck	0	5	0	9	,, 13	Double ditto ditto	0	6	1	0
,, 6	Double ditto with Neck	0	10	1	3	,, 14	Single Dipper, with Rim	0	8	1	0
,, 7	Single ditto with Screw Top	0	8	1	4	,, 15	Double ditto ditto	1	4	2	0
,, 8	Single ditto with Capped Lid	0	9								

No. 66.—OVAL AND ROUND TIN CASES,
FOR PENCILS, ETC.

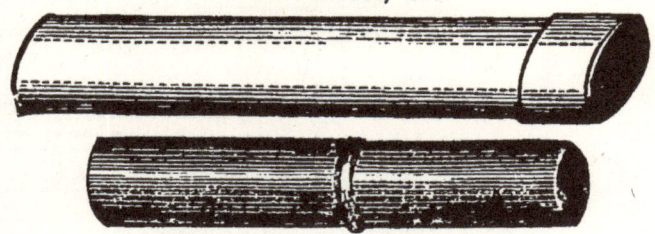

4d., 6d., 1/-, 1/3, 1/6, 1/9, and 2/- each.

No. 67.—WRITERS' AND DECORATORS' WOOD COMPASSES,
VARIOUS PATTERNS.

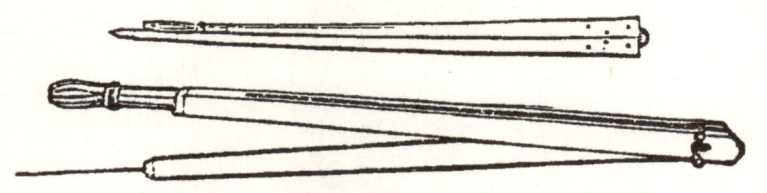

1/3, 1/6, 1/9, 2/-, 2/3, 2/6, 3/-, 3/6, and 4/- each.

No. 68.—WRITERS' GOLD SIZE.

9d. per bottle, or 4/- per quart.

No. 69.—OIL GOLD SIZE.

Best Oil Gold Size kept ready for use, and thick in pots, 1-oz., 2-oz., 4-oz., 6-oz., 8-oz., 12-oz., *and* 16-oz.

3d., 4½d., 7½d., 1/-, 1/3, 2/-, and 2/6 each.

Matt Size, 1/- lb.   Burnish Size, 1/- lb.

No. 70.—COTTON WOOL, 2/6 per lb.

No. 71.—PARCHMENT CUTTINGS, 1/- per lb.

No. 72.—PIPE CLAY, in Sticks, 3d. per doz.

No. 73.—TISSUE PAPER, 6d. per Quire.

No. 74.—GOLD LEAF.

Deep Full Size, 1/2 per book, 45/- thousand.
Pale ditto   1/1   ,,   42/6   ,,
Half-gold ditto 5d.   ,,   12/6   ,,
Dutch Metal   2d.   ,,
Transfer Gold   1/4   ,,

No. 75.—CAMEL-HAIR PENCILS, SHORT.

CROW.

DUCK.

GOOSE.

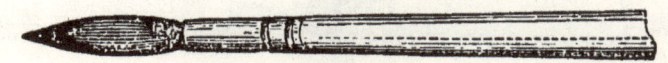

FULL GOOSE.

Crow, 8/-; Duck, 10/-; Goose, 12/-; Full Goose, 14/-
Assorted, G. D. C., 10/- gross.

No. 76.—CAMEL-HAIR, MIDDLE.

CROW.

DUCK.

GOOSE.

Crow, 8/-; Duck, 10/-; Goose, 12/-. Assorted, 10/- gross.

No. 77.—CAMEL-HAIR WRITERS.

CROW.

DUCK.

GOOSE.

Crow, 8/-; Duck, 10/-; Goose, 12/-. Assorted, 10/- gross.

No. 78.—CAMEL-HAIR JAPANNERS' LONG WRITERS.

CROW.

DUCK.

GOOSE.

Crow, 8/-; Duck, 10/-; Goose, 12/-; Assorted, 10/- gross.

No. 79.—CAMEL-HAIR SWAN PENCILS.

Small, 18/-; Large, 24/-.

No. 80.—CAMEL-HAIR SWAN TINS.

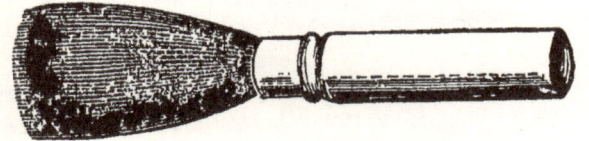

24/- gross. Ex. Large, 42/-

No. 81.—CAMEL-HAIR LINERS, OR STRIPERS.

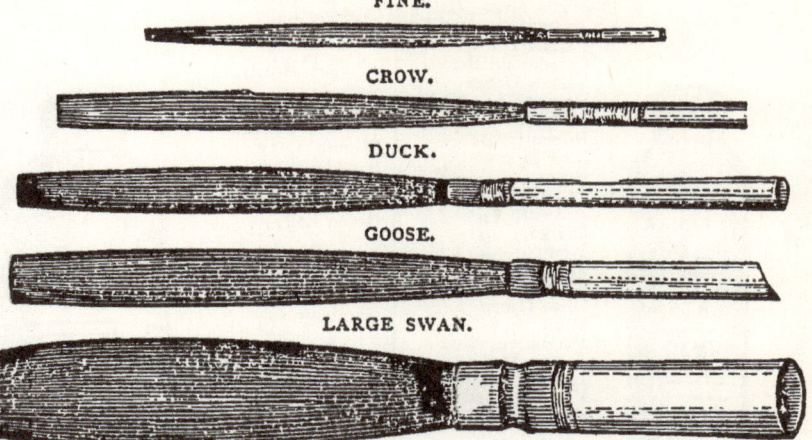

FINE.
CROW.
DUCK.
GOOSE.
LARGE SWAN.

Fine, 8/-; Crow, 8/-; Duck, 10/-; Goose, 12/-; Swan, 24/-;
Middle Swan, 36/-; Large Swan, 48/- gross.

No. 82.—WRITERS' BOXES, &c., &c.

Boxes for Sign Writers, Glass Embossers, Decorators, Missal Painters, Gilders, &c.,

Price 21/-

BOXES MADE TO ANY DESIGN.

Containing Boxes with Lids, for Colors, &c.; Screw Bottles, for Varnishes, Oils, &c.; Spaces for Gold Leaf Brushes, Tube Colors, and Palette.

OIL COLORS.

FINELY GROUND BY MACHINERY. IN PATENT COLLAPSIBLE TUBES.

Single, 3d.
Double Size, 6d.
Larger, 9d. and 1s.

Verona Brown
Caledonian Brown
Yellow Ochre
Roman or Golden Ochre
Brown Ochre
Raw Sienna
Burnt ,,
Prussian Blue
Antwerp ,,
Indigo ,,
New ,,
Magenta
Mauve
Olive Green
Sap ,,
Permanent Blue
Indian Red
Light ,,
Venetian Red
Bone Brown
Cappah ,,
Vandyke ,,
Pale Naples Yellow
Deep ,, ,,
Extra Deep Naples Yellow
Patent Yellow
Chromes, Nos. 1, 2, 3, 4
Raw Umber
Burnt ,,
Sacrum, or Dryers
Mineral Green
Mummy
Bitumen
Asphaltum
Dutch Pink
Brown ,,
Italian ,,
Neutral Tint
Red Lead
Cologne Earth
Yellow Lake
Purple ,,
Indian ,,
Lake
Ivory Black
Lamp ,,
Blue ,,
Terravert
Verdigris
Emerald Green
Flake White
Cassell Earth
Blue Verditer
White Zinc

4d. each.
Crimson Lake
Scarlet ,

5d. each.
Vermilion

6d. each.
Mineral Grey
Royal Yellow
Cerulean (blue)
French Ultramarine
Green Lake
,, No. 1
,, No. 2
,, No. 3
,, No. 4
Gamboge

8d. each.
Oxide of Chromium
Mars Yellow
Strontian Yellow
Indian ,,

9d. each.
Lemon Yellow, Pale
,, ,, Deep
Cobalt
Scarlet Vermilion Ext. ,,

10d. each.
Madder Brown
Mars Orange
Sepia

1s. each.
Pink Madder
Madder Lake
Malachite Green
Rose Madder
Pale Ultramarine Ash
Veronese Green

1s. 6d. each.
Carmine
Violet Carmine
Cadmium Yellow
Deep Ultramarine Ash
Purple Madder
Madder Carmine

2s. 6d. each.
Aureolin

SINGLE. DOUBLE.

WHITE { Double 6d. Treble 8d. Quadruple 1s. 1-lb. 2s. 6d. 2-lb. 4s. 6d. each.

MYGULPH (Mastic) { Single 3d. Double 6d.

MYGULPH (Copal) { Single 3d. Double 6d.

Chrome Green, Light, Middle, Deep, 3d. each.

GILDERS' BRUSHES, &c.

No. 83.—GILDERS' DABBERS OR MOPS IN QUILLS.
Wire-bound. Flat on Top or Domed.

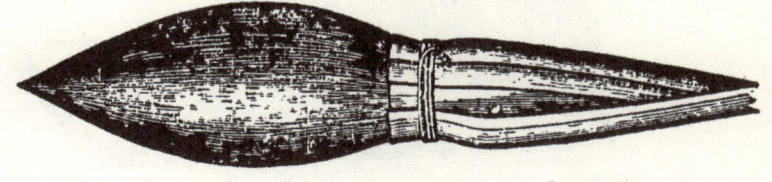

1	2	3	4	5	6 Quill.
2/6	3/6	4/6	6/-	8/-	9/- dozen.

No. 84.—GILDERS' MOPS ON HANDLES.
Flat on Top or Domed.

No. 1	2	3	4	5	6	7	8	9	10	11	12
3/6	4/6	6/-	7/6	9/-	10/6	13/-	15/-	17/-	20/-	22/-	24/-doz

No. 85.—GILDERS' TIPS.
THREE DIFFERENT LENGTHS OF HAIR.

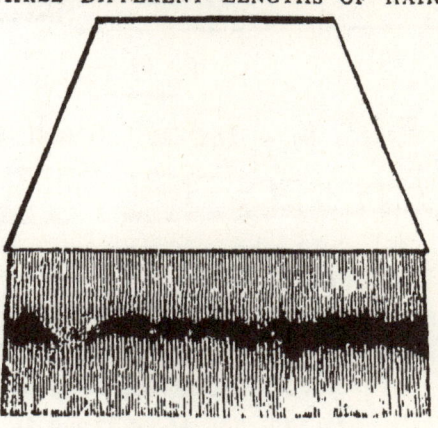

1 inch, 1½ inch, 2 inches long, and 4 inches wide—3/- dozen.

No. 86.—GILDERS' CUSHIONS.

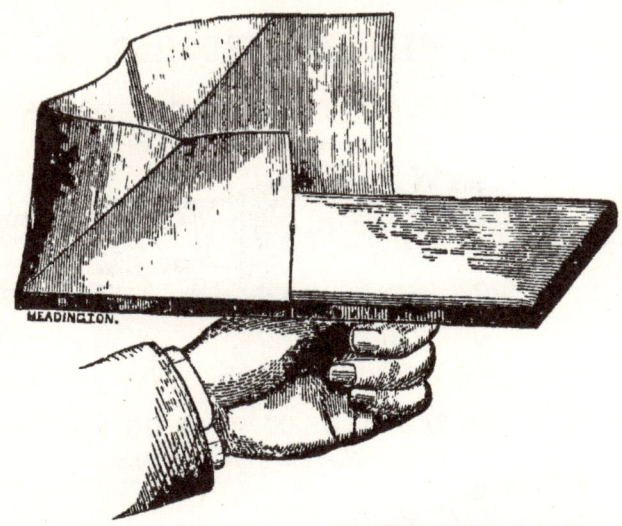

8¼ × 5½ Size, 15/- dozen. 9¼ × 6½ Size, 18/- dozen.

No. 87.—GILDERS' KNIVES, BALANCE HANDLES.

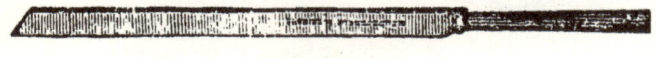

12/- dozen.

No. 88.—GILDERS' KNIVES.

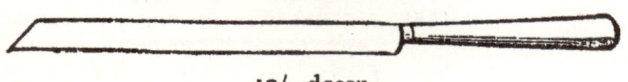

10/- dozen.

No. 89.—GILDERS' BURNISHERS.

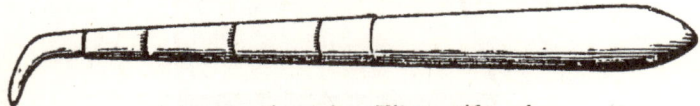

Agate, 2/- each; Flint, 2/6 each.

No. 90.—LONG HANDLE BURNISHER,
FOR BOOK BINDERS, &c.

2/6, 3/-, 3/6, 4/-, 4/6, and 5/- each.

No. 91.—QUILLED TOOLS, OR SKEWING BRUSHES.

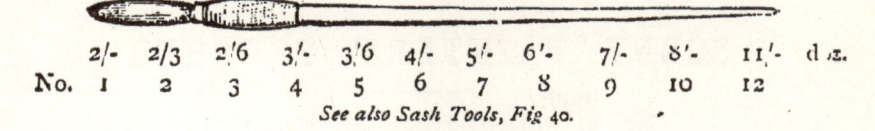

No.	2/-	2/3	2/6	3/-	3/6	4/-	5/-	6/-	7/-	8/-	11/-	d.z.
	1	2	3	4	5	6	7	8	9	10	12	

See also Sash Tools, Fig 40.

No. 92.—ROUND FRENCH TOOLS,
STRING BOUND, FINE HAIR.

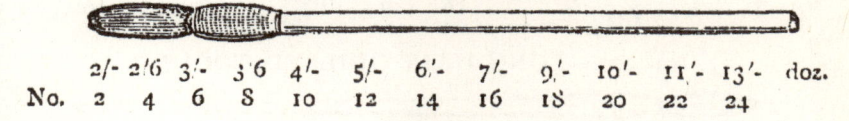

No.	2/-	2/6	3/-	3/6	4/-	5/-	6/-	7/-	9/-	10/-	11/-	13/-	doz.
	2	4	6	8	10	12	14	16	18	20	22	24	

No. 93.—OIL GOLD SIZE,

Thick and ready for use, prepared in Pots. See Fig 69.

3d., 4½d., 7½d., 1/-, 1/3, 2/-, and 2/6 each.

No. 94.—MATT AND BURNISH GOLD SIZE, 1/- per lb.

No. 95.—GILDERS' COTTON, 2/6 per lb.

No. 96.—PARCHMENT CUTTINGS, 1/- per lb.

No. 97.—GOLD LEAF,
See Fig 70.

Deep full size, 1/2 per book.
Ditto, pale, 1/1 ,,

No. 98.—CANING TUBES,
FOR CARRIAGE BUILDERS, &c.

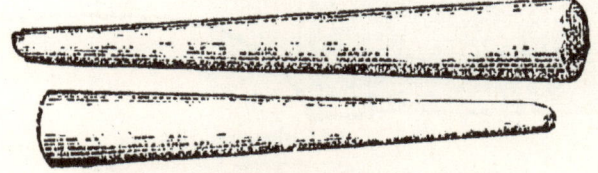

Single, 1/6 each; Double, 2/6 each.

SCENE PAINTERS' BRUSHES.

No. 99.—LONG HAIR SASH TOOL.

No.	3	4	5	6	7	8	9	10	12	14	16	
	8/-	10/-	13/-	16/-	18/-	20/-	24/-	30/-	34/-	39/-	48/-	doz.

No. 100.—LONG HAIR QUILLED TOOLS.

No.	1	2	3	4	5	6	7	8	9	10	11	12	
	2/6	3/-	3/6	4/-	5/-	6/-	7/-	8/-	9/-	10/-	11/-	12/-	doz.

No. 101.—SHORT HAIR QUILLED TOOLS.

No.	1	2	3	4	5	6	7	8	9	10	11	12	
	2/-	2/6	3/-	3/6	4/6	5/6	7/-	8/-	9/-	10/-	11/-	12/-	doz.

No. 102.—ONE KNOT DISTEMPER.

7-inch Hair, 4/- and 4/6 each.   6-inch Hair, 3/-, 3/6, 4/- each.

No. 103.—FLAT COPPER BAND DISTEMPERS.

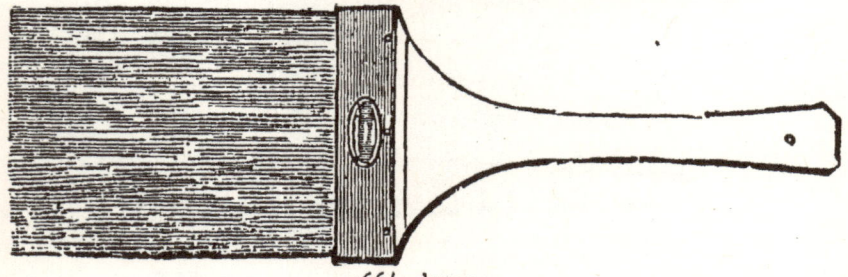

66/- dozen.

No. 104.—TWO KNOT GROUND DISTEMPERS.
(See Fig 39.)

Messrs. BRODIE & MIDDLETON are prepared to supply Amateurs with Colors prepared for Scene Painting in small or large quantities.

		s.	d.				s.	d.
White Lead	per lb.	1	0	Maroon		per lb.	2	0
No. 1 Chrome	,,	1	0	Purple		,,	2	0
No. 2 Chrome	,,	1	0	Light Blue		,,	2	0
No. 3 Chrome	,,	1	0	Dark Blue		,,	3	6
Dutch Pink	,,	0	9	Ultramarine		,,	2	6
Brown Ochre	,,	0	8	Emerald Green		,,	2	6
Venetian Red	,,	0	8	Dark Green		,,	2	0
Orange Red	,,	1	0	Raw Sienna		,,	1	6
Persian Red	,,	1	9	Burnt Sienna		,,	1	6
Magenta	,,	1	9	Vandyke Brown		,,	1	6
Damp Lake	,,	4	0	Ivory Drop Black		,,	1	6
Carnation	,,	5	6					

Tans to hold 1 lb. of the above, 2d. each.

Theatrical Canvas, of all sizes, prepared for Scene Painting, 4d. per Superficial Square Foot.

No. 105.—MADONG, 2/6 per lb.

EVERY REQUISITE PROVIDED FOR SCENE PAINTERS

VARNISH BRUSHES AND TOOLS,
WARRANTED TO STAND.

No. 106.—OVAL VARNISH BRUSHES,
TIN BOUND.

No.	1	2	3	4	5	6	7	8
	15/-	18/-	24/-	27/-	30/-	36/-	42/-	48/- doz.

No. 107.—CARRIAGE VARNISH BRUSHES.

No.	1	2	3	4	5	6	7	8
	15/-	18/-	24/-	27/-	30/-	36/-	42/-	48/- doz.

No. 108.—OVAL COPPER-WIRE BOUND VARNISH BRUSHES.

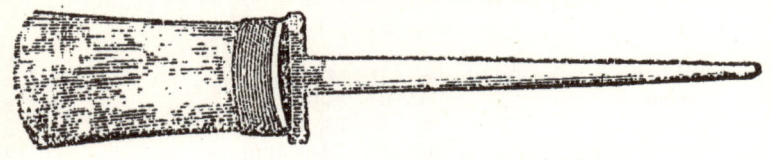

15/-	24/-	30/-	45/-	54/- doz.
No. 1	2	3	4	5

No. 109.—VARNISH TOOLS,

In Socket, and Wire Bound.

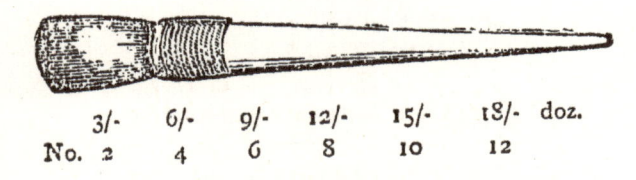

3/-	6/-	9/-	12/-	15/-	18/- doz.
No. 2	4	6	8	10	12

No. 110.—PINNED VARNISH BRUSHES.

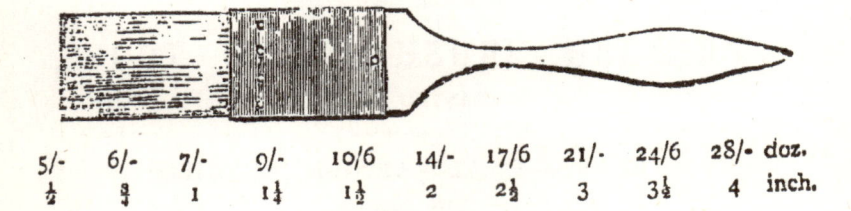

5/-	6/-	7/-	9/-	10/6	14/-	17/6	21/-	24/6	28/- doz.
½	¾	1	1¼	1½	2	2½	3	3½	4 inch.

No. 111.—BEARS' HAIR VARNISH BRUSHES.

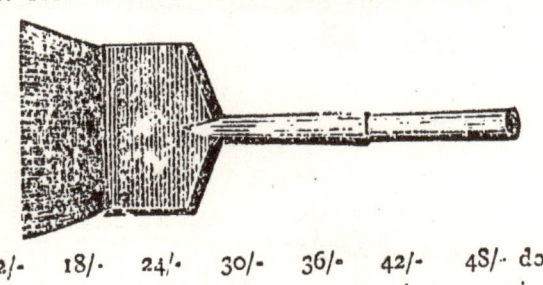

12/-	18/-	24/-	30/-	36/-	42/-	48/- doz.
1	1½	2	2½	3	3½	4 inch.

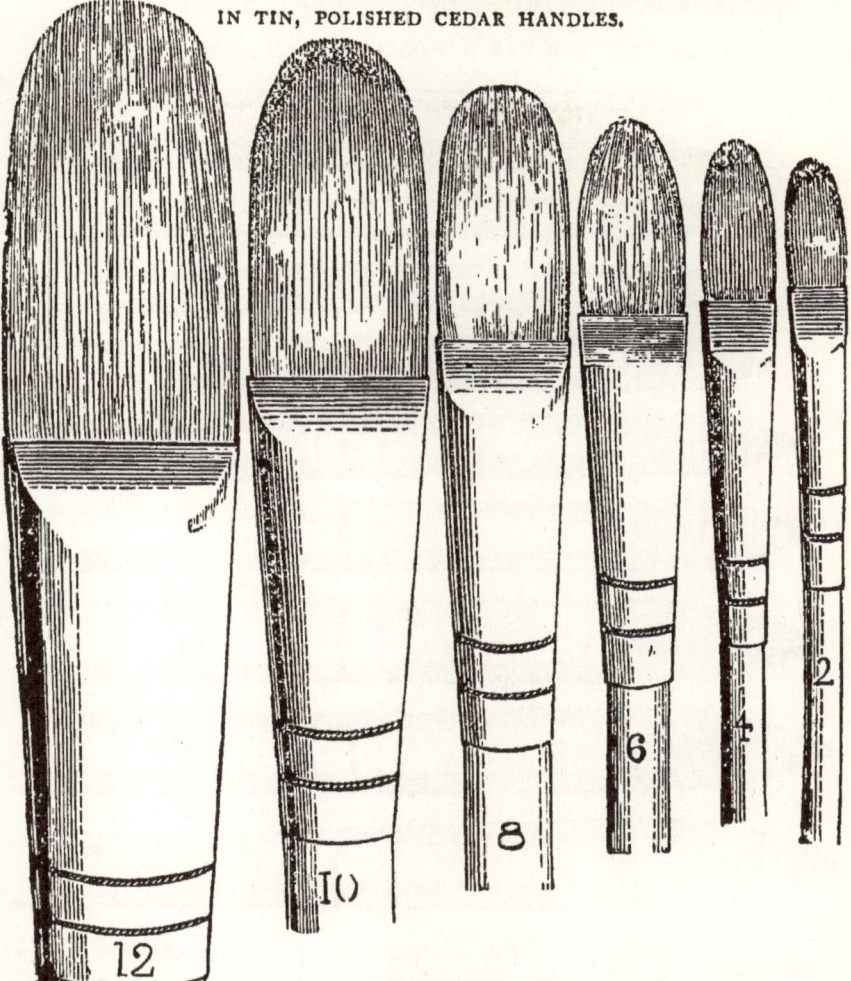

No. 112.—ROUND AND FLAT HOG-HAIR TOOLS,
IN TIN, POLISHED CEDAR HANDLES.

Flat or Round same price.

No.	1	2	3	4	5	6	7	8	9	10	11	12
	2/6	2/6	3/-	3/-	3/6	4/-	4/6	5/-	6/-	7/-	8/-	8/- doz.

No.	13	14	16	18	20	22	24
	9/-	10/-	12/-	14/-	16/-	18/-	21/- doz.

No. 113.—ROUND AND FLAT FRENCH TOOLS,
IN TIN—WHITE HANDLES—ENGLISH SIZES.

No.	1	2	3	4	5	6	7	8	9	10	11	12
	2/-	2/-	2/-	2/6	2/6	3/-	3/-	3/6	4/-	5/-	6/-	6/- doz.

No. 114.—SABLE TOOLS, IN TIN, FLAT AND ROUND,
POLISHED CEDAR HANDLES, 12-INCH LONG.

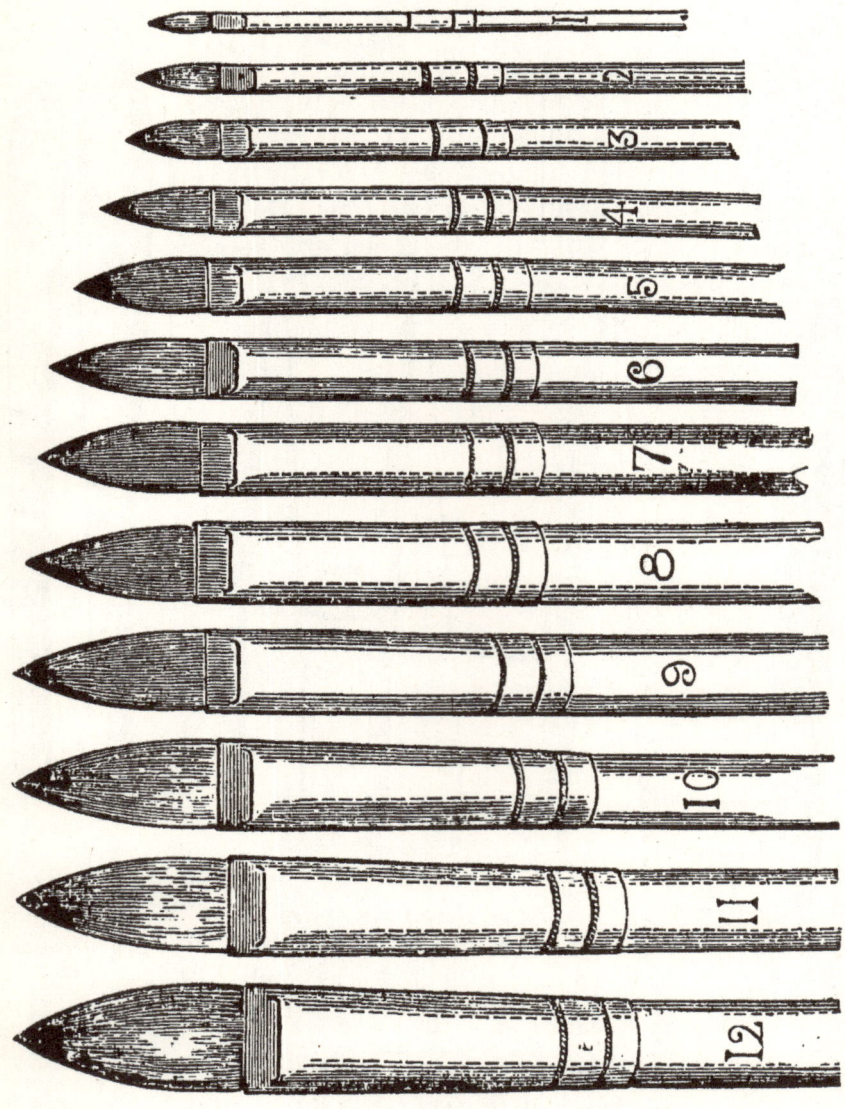

Flat	4/6	5/-	6/-	7/-	8/-	9/-	12/-	16/-	20/-	24/-	30/-	36/- doz
Round	4/6	5/-	6/-	7/-	8/-	9/-	16/-	20/-	22/-	33/-	42/-	51/- ,,
No.	1	2	3	4	5	6	7	8	9	10	11	12

MISCELLANEOUS.

No. 115.—FLAT C. HAIR VARNISH BRUSHES,
In Tin—Best quality.

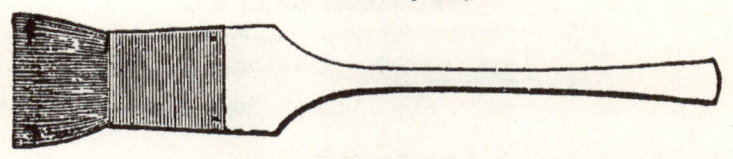

THICKNESS	inch.	½	¾	1	1¼	1½	1¾	2	2½	3	3½	4	inch.
	⅛ ...	4/-	5/-	6/-	7/6	9/-	10/6	12/-	15/-	18/-	21/-	24/-	dozen.
	3/16 ...	5/-	6/-	8/-	—	12/-	—	16/-	20/-	24/-	28/-	32/-	,,
	¼ ...	—	—	10/-	—	15/-	—	20/-	25/-	30/-	35/-	40/-	,,
	5/16 ...	—	—	—	—	—	—	22/-	28/-	33/-	39/-	44/-	,,
	⅜ ...	—	—	—	—	—	—	24/-	—	36/-	—	48/-	,,
	½ ...	—	—	—	—	—	—	30/-	—	45/-	—	60/-	,,

No. 116.—ROUND C. HAIR VARNISH OR LACQUERING BRUSHES.

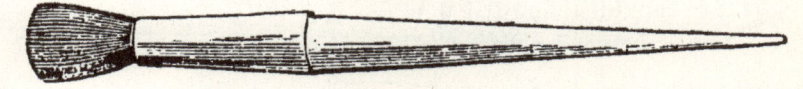

2/-	2/6	3/-	4/-	5/-	6/-	7/-	8/-	9/-	11/-	12/-	13/-	doz.
No. 1	2	3	4	5	6	7	8	9	10	11	12	

Larger sizes to order.

No. 117.—FLAT BADGER SOFTENERS,
IN TIN SOCKETS.

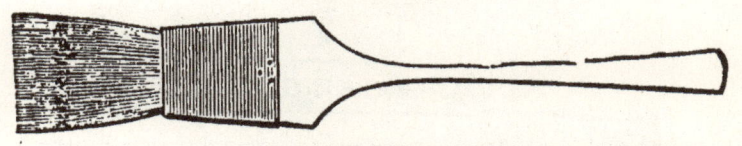

8/-	12/-	16/-	20/-	24/-	32/-	36/-	48/-	56/-	64/-	doz.
½	¾	1	1¼	1½	2	2¼	3	3½	4	inch.

No. 118.—ROUND BADGER SOFTENERS.

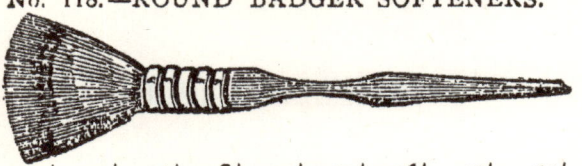

No.	1	2	3	4	5	6	7	8	9	10	11	12
	6/-	8/-	10/-	12/-	15/-	18/-	24/-	30/-	36/-	45/-	54/-	60/- doz.

No. 119.—BORDERING BRUSH.

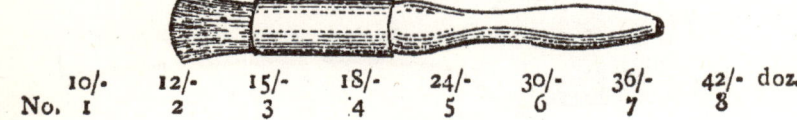

No.	1	2	3	4	5	6	7	8
	10/-	12/-	15/-	18/-	24/-	30/-	36/-	42/- doz.

No. 120.—POONAH BRUSH,
POLISHED CEDAR HANDLES.

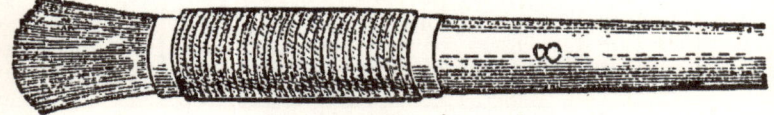

1 to 6 Assorted, 3/- per doz.

No. 121.—POONAH, OR VELVET SCRUB,
POLISHED CEDAR HANDLES.

1 to 6 Assorted, 3/- per doz.

No. 122.—BADGER POONAH,
POLISHED CEDAR HANDLES.

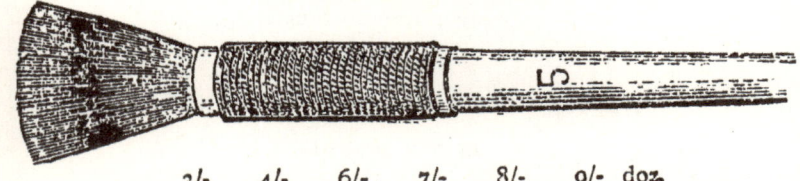

No.	1	2	3	4	5	6
	3/-	4/-	6/-	7/-	8/-	9/- doz.

No. 123.—MARKING BRUSH.

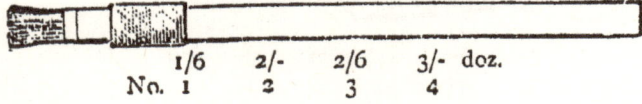

No.	1	2	3	4
	1/6	2/-	2/6	3/- doz.

Printed in September 2023
by Rotomail Italia S.p.A., Vignate (MI) - Italy